Ion Channels

METHODS IN MOLECULAR BIOLOGY™

John M. Walker, SERIES EDITOR

10 0483152 9

METHODS IN MOLECULAR BIOLOGY™

Ion Channels

Methods and Protocols

Edited by

James D. Stockand
Mark S. Shapiro

University of Texas Health Science Center
San Antonio, TX

HUMANA PRESS ✳ TOTOWA, NEW JERSEY

Production Editor: Jennifer Hackworth

Cover design by Patricia F. Cleary

Cover illustration: Figure 2 from Chapter 1, "Functional Reconstitution of the Human Epithelial Na⁺ Channel in a Mammalian Expression System," by Alexander Staruschenko, Rachell E. Booth, Oleh Pochynyuk, James D. Stockand, and Qiusheng Tong.

For additional copies, pricing for bulk purchases, and/or information about other Humana titles, contact Humana at the above address or at any of the following numbers: Tel.: 973-256-1699; Fax: 973-256-8341; E-mail: orders@humanapr.com; or visit our Website: www.humanapress.com

Printed in the United States of America. 10 9 8 7 6 5 4 3 2 1
1-59745-095-2 (e-book)
ISSN 1064-3745   ℓ004831529

Library of Congress Cataloging-in-Publication Data

Ion channels : methods and protocols / edited by James D. Stockand, Mark S. Shapiro.

  p. ; cm. -- (Methods in molecular biology, ISSN 1064-3745 ; v. 337)

  Includes bibliographical references and index.

  ISBN 1-58829-576-1 (alk. paper)

  1. Ion channels.

  [DNLM: 1. Ion Channels. QU 55 I69 2006] I. Stockand, James D.

II. Shapiro, Mark S. III. Series: Methods in molecular biology (Clifton, N.J.) ; v. 337.

  QH603.I54I59 2006

  571.6'4--dc22   2005026667

Acknowledgment

Ms. Lea Harlow is acknowledged for excellent copyediting support.

Preface

Research on ion channels has exploded in the last few decades and it is now clear that ion channels play essential roles in cell biology and physiology, with their dysfunction being the root cause of many human diseases. Understanding human biology in the post-genome sequencing era requires that the function of the protein products encoded by these recently sequenced genes be quantified. Using contemporary tools and new experimental approaches, scientists interested in ion channels are in the unique position of being able to directly, and often in real-time, measure ion channel activity, subunit stoichiometry, structure–function relationships, as well as many other biophysical/biochemical parameters regarding a channel of interest. Development and application of these experimental tools has led to a boom in investigation of ion channels; however, as in many fields of research, wide implementation of the newest technological advances lags well behind the discovery of these advances. We believe that recently developed technologies useful for studying ion channels have matured enough to where they should now be readily available to any interested scientist. *Ion Channels* provides a comprehensive and detailed description of recent technological breakthroughs and experimental designs used to successfully study ion channels.

We write *Ion Channels* with the hopes that it will provide insight into rational experimental design and the practical application of methodologies for research on ion channels. The book is designed as a guide to facilitate emerging scientists and young investigators beginning to establish their independent laboratories. In addition, the established investigator whose research has recently directed them towards the study of ion channels will find the book useful. This edition also details several recent methodological breakthroughs with respect to study of ion channels that the established ion channel investigator will find useful. Finally, this edition will serve as a general and hopefully handy resource with respect to practical application of diverse experimental approaches for studying ion channels. It is our goal that this edition will provide scientists the wherewithal to implement new research strategies and methods into their research programs.

Ion Channels is mainly about practical implementation of the study of ion channels and should be used as a field guide by the investigator when designing and performing research on channels. Each chapter provides detail about a particular aspect of investigating ion channels. This detail includes description of the actual successfully employed experimental procedures and pitfalls for each application. The applications covered here are broad-ranging, from the study of allosteric regulation of ion channel activity using a classic mutagenesis approach to the study of channel subunit stoichiometry using a novel biophysical approach based on fluorescence resonance energy transfer. We feel that our book is both comprehensive and practical providing important information that every scientist interested in ion channels should know.

James D. Stockand
Mark S. Shapiro

Contents

Contributors

RICHARD M. AHLQUIST • *Department of Physiology and Biophysics, University of Washington, Seattle, WA*

MOUHAMED S. AWAYDA • *Department of Physiology and Biophysics, SUNY at Buffalo, Buffalo, NY*

ISABELLE BARÓ • *L'institut du Thorax, Inserm U533, Université de Nantes, Nantes, France*

FRIEDRICH BEERMANN • *Mutant Mouse Core Facility, ISREC, Epalinges, Switzerland*

ABDERRAHMANE BENGRINE • *Department of Physiology and Biophysics, SUNY at Buffalo, Buffalo, NY*

DALE J. BENOS • *Department of Physiology and Biophysics, University of Alabama at Birmingham, Birmingham, AL*

BAKHROM K. BERDIEV • *Department of Physiology and Biophysics, University of Alabama at Birmingham, Birmingham, AL*

RACHELL E. BOOTH • *Department of Chemistry and Biochemistry, Texas State University, San Marcos, TX*

COSTA M. COLBERT • *Department of Biology and Biochemistry, University of Houston, Houston, TX*

LYNETTE C. DAWS • *Department of Physiology, University of Texas Health Science Center, San Antonio, TX*

CAROL DEUTSCH • *Department of Physiology, University of Pennsylvania, Philadelphia, PA*

DENIS ESCANDE • *L'institut du Thorax, Inserm U533, Université de Nantes, Nantes, France*

NIKITA GAMPER • *School of Biomedical Sciences, University of Leeds, Leeds, UK*

SABRINA GUICHARD • *Mutant Mouse Core Facility, ISREC, Epalinges, Switzerland*

CHOU-LONG HUANG • *Department of Medicine, UT Southwestern Medical Center, Dallas, TX*

EDITH HUMMLER • *Department of Pharmacology and Toxicology, Transgenic Animal Facility, Faculty of Biology and Medicine and the University Hospitals, University of Lausanne, Lausanne, Switzerland*

THOMAS R. KLEYMAN • *Departments of Medicine and Cell Biology and Physiology, University of Pittsburgh, Pittsburgh, PA*

DUK-SU KOH • *Department of Physiology and Biophysics, University of Washington, Seattle, WA, Department of Physics, Pohang University of Science and Technology, Pohang, Republic of Korea*

FARHAD KOSARI • *Department of Health Sciences Research, Mayo Clinic, Rochester, MN*

ANDREY KOSOLAPOV • *Department of Physiology, University of Pennsylvania, Philadelphia, PA*

GILDAS LOUSSOUARN • *L'institut du Thorax, Inserm U533, Université de Nantes, Nantes, France*

ANNE-MARIE MÉRILLAT • *Department of Pharmacology and Toxicology, University of Lausanne, Lausanne, Switzerland*

LYUBOV V. PARFENOVA • *Department of Physiology, University of Texas Health Science Center, San Antonio, TX*

OLEH POCHYNYUK • *Department of Physiology, University of Texas Health Science Center at San Antonio, San Antonio, TX*

ANDRÉE PORRET • *Transgenic Animal Facility, Faculty of Biology and Medicine and the University Hospitals, University of Lausanne, Lausanne, Switzerland*

JOHN M. ROBINSON • *Department of Physiology, University of Pennsylvania, Philadelphia, PA*

BRAD S. ROTHBERG • *Department of Physiology, University of Texas Health Science Center, San Antonio, TX*

WEIJIAN SHAO • *Department of Physiology and Biophysics, SUNY at Buffalo, Buffalo, NY*

MARK S. SHAPIRO • *Department of Physiology, University of Texas Health Science Center at San Antonio, San Antonio, TX*

SHAOHU SHENG • *Department of Medicine, University of Pittsburgh, Pittsburgh, PA*

ALEXANDER STARUSCHENKO • *Department of Physiology, University of Texas Health Science Center at San Antonio, San Antonio, TX*

JAMES D. STOCKAND • *Department of Physiology, University of Texas Health Science Center at San Antonio, San Antonio, TX*

JANE M. SULLIVAN • *Department of Physiology and Biophysics, University of Washington, Seattle, WA*

GLENN M. TONEY • *Department of Physiology, University of Texas Health Science Center, San Antonio, TX*

QIUSHENG TONG • *Department of Physiology, University of Texas Health Science Center at San Antonio, San Antonio, TX*

IVANA VUKOJICIC • *Department of Physiology and Biophysics, SUNY at Buffalo, Buffalo, NY*

JIE ZHENG • *Department of Physiology and Membrane Biology, University of California at Davis, School of Medicine, Davis, CA*

I

Methods for Exogenous Expression of Ion Channels in Cells

1

Functional Reconstitution of the Human Epithelial Na⁺ Channel in a Mammalian Expression System

Alexander Staruschenko, Rachell E. Booth, Oleh Pochynyuk, James D. Stockand, and Qiusheng Tong

Summary

Probing ion channel structure–function and regulation in native tissue can, in some instances, be experimentally challenging or impractical. To facilitate discovery and increase experimental flexibility, our laboratory routinely reconstitutes recombinant ion channels in a mammalian expression system quantifying channel activity with patch clamp electrophysiology. Here, we describe investigation of the human epithelial Na⁺ channel heterologously expressed in Chinese hamster ovary cells.

Key Words: Heterologous expression; ion channel; patch clamping; transient transfection.

1. Introduction

Ion channels serve as cellular gateways capable of changing electrical signals to chemical signals and vice-versa. The patch clamp technique enables direct measurement of ion channel function (*1*). The marriage of contemporary molecular biology and genetics to identify channel mutations in humans with functional results from patch clamp recordings of mutant channels has defined the cellular and molecular underpinnings of many inherited diseases, including cystic fibrosis and some forms of congenital hypertension, muscular dystrophy, epilepsy, febrile seizures, ataxia, deafness, myotonia, and heart arrhythmias (*2*).

In many cases, it is impractical to study ion channels in native tissues and preparations. This is particularly true regarding the study of structure–function relationships, and some forms of channel regulation for ion channels are often found in the plasma membrane of native cells at low frequencies, making their

From: *Methods in Molecular Biology, vol. 337: Ion Channels: Methods and Protocols*
Edited by: J. D. Stockand and M. S. Shapiro © Humana Press Inc., Totowa, NJ

investigation extremely labor intensive or even inconsequential when channel activity is lost in the noise of the system. Reliable native preparations for ion channels found in inaccessible tissues or tissue extremely sensitive to their surroundings are also not always available. In many cases, native preparations do not allow enough experimental flexibility. To surmount these limitations, investigators have begun overexpressing ion channels of interest in heterologous expression systems.

Improvement in cell culture technologies and the molecular cloning of ion channel genes have greatly facilitated the overexpression of recombinant ion channels in expression systems. It is now routine to study ion channels in precisely controlled experiments in cells with a uniform set of properties and attributes with favorable signal-to-noise ratios. This allows investigation of the structure–function relation and regulation of virtually any ion channel for which the molecular correlate is known. Here, we describe expression of epithelial Na^+ channels (ENaCs; reviewed in **refs. 3–5**) in Chinese hamster ovary (CHO) cells with recombinant channel activity assayed using the patch clamp method.

2. Materials

2.1. Overexpression of ENaCs in CHO Cells

1. CHO cells (American Tissue Culture Collection, Manassas, VA, CHO-K1 CCL-61).
2. Dulbecco's modified Eagle's medium (DMEM).
3. Fetal bovine serum (FBS).
4. 10 mg/mL Streptomycin/10,000 IU/mL penicillin stock solution.
5. 0.05% Trypsin/0.53 mM ethylenediaminetetraacetic acid stock solution.
6. Phosphate-buffered saline (PBS) without $CaCl_2$ (10X stock).
7. Polyfect reagent (Qiagen, Valencia, CA, 301107).
8. pEGFP-F (Clontech, cat. no. 632308) used to identify transfected cells.
9. Purified channel cDNA in an appropriate expression plasmid containing a mammalian promoter (*see* **Notes 1–3**). We used human α-ENaC, β-ENaC, and γ-ENaC in pMT3 and mouse α-ENaC, β-ENaC, and γ-ENaC in pCMV-Myc (BD Biosciences Clontech, Palo Alto, CA). Creation of these constructs is described in **refs. 6–8**.
10. 0.01% Polylysine solution (10X stock).
11. 18 × 18 – 2 Coverglass (Fisher Scientific, Pittsburgh, PA, 12-540-A) cut into chips 5 × 5 mm.
12. 10 mM Amiloride in dimethyl sulfoxide (1000X stock).

2.2. Analysis of Recombinant ENaCs Using the Patch Clamp Method

1. Axopatch 200B patch clamp amplifier (Axon Instruments, Union City, CA; *see* **Note 4**).

2. Digidata 1322A analog-to-digital board (Axon Instruments) interfaced with a personal computer running the *pClamp 9.2* software suite (Axon Instruments).
3. MP-285 micromanipulator (Sutter Instrument Co., Novato, CA).
4. Microvibration isolation table with Faraday cage (Technical Manufacturing Co., Peabody, MA) floating on nitrogen.
5. TE2000-U inverted microscope (Nikon, Melville, NY) fitted with epifluorescence and excitation and emissions filters for EGFP (FITC/RSGFP LP emission; 21012, Chroma Technical Corp., Rockingham, VT).
6. Model P-97 flaming/brown micropipet puller (Sutter Instrument Co.).
7. MF-830 microforge (Narishige, East Meadow, NY).
8. Borosilicate glass capillaries (1B150F-4, World Precision Instruments, Sarasota, FL) pulled and forged to 5–6 and 2–3 mΩ for excised and whole-cell patch recording, respectively.
9. Fast exchange recording chamber (model RC-22, Warner Instruments, Hamden, CT).
10. Valve Bank II with pinch valve perfusion system (AutoMate Scientific Inc., San Francisco, CA).
11. Intracellular pipet solution: 140 mM CsCl, 2 mM adenosine triphosphate, 2 mM MgCl$_2$, 5 mM ethyleneglycol-bis(beta-aminoethyl ether)-N,N,N',N'-tetraacetic acid (EGTA), 0.1 mM guanosine 5'-triphosphate, 5 mM NaCl, 10 mM HEPES, pH 7.4 (*see* **Note 5**).
12. Extracellular bathing solution: 160 mM NaCl, 1 mM CaCl$_2$, 2 mM MgCl$_2$, 10 mM HEPES, pH 7.4.

3. Methods

The CHO cell line initiated by Puck and colleagues (*9*) is an excellent expression system for reconstituting ENaCs (*see* **Note 6**). These cells (1) have been immortalized; (2) have homogeny in size, shape, and other intrinsic properties; (3) efficiently express exogenous cDNA from many different mammalian promoters; (4) provide ready access to the patch pipet; (5) readily form high-resistance (gigaohm) seals; and (6) have low background currents (little endogenous ion channel activity), with none sensitive to amiloride (*6,7,10–13*).

Overexpression of a channel, such as an ENaC, allowing a constitutive inward Na$^+$ current into a nonpolar CHO cell is lethal because of (1) continued development of osmotic disequilibrium caused by a massive influx of Na$^+$ and (2) constant and inappropriate membrane depolarization. **Figure 1** shows the resting membrane potential of an untransfected CHO cell (top) and a CHO-expressing mouse ENaC (bottom) in the presence and absence of a channel blocker (10 μM amiloride; **ref. 3**). The presence of active ENaCs strongly depolarizes the membrane toward the Nernst equilibrium for Na$^+$. Maintaining CHO cells overexpressing ENaCs with 10 μM amiloride until just prior to experimentation counters inappropriate membrane depolarization and thus circumvents the need to modify tissue culture media and patch solutions.

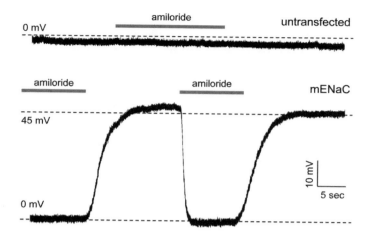

Fig. 1. Resting membrane potential in untransfected Chinese hamster ovary (CHO) cells and cells overexpressing recombinant epithelial Na^+ channels (ENaC). Shown here are resting membrane potentials in physiological saline of a current clamped, untransfected CHO cell (top) and a cell overexpressing recombinant mouse ENaC (mENaC; bottom) before and after amiloride. Current clamps were formed without disruption of the cytosol using the perforated patch method.

3.1. Overexpression of ENaCs in CHO Cells

1. All tissue culture and transfection is performed using sterile techniques in a tissue culture hood with sterile solutions and disposables.
2. CHO cell stocks are maintained as subconfluent adherent cultures in 100-mm tissue culture dishes in DMEM supplemented with 10% FBS and 5% streptomycin/penicillin in a humidified tissue culture incubator at 37°C and 5% CO_2. Stock cell medium is replaced every other day.
3. Stock cells are passed twice a week when they are approx 85% confluent with 10–20% of the passed cells used to continue the stock with the remainder used for transfection and subsequent experimentation.
4. Cells to be passed are released from the 100-mm tissue culture dish using 1 mL trypsin/ethylenediaminetetraacetic acid for 10–15 min following a wash with PBS (*see* **Note 7**). After release, trypsin is inactivated with 3 mL tissue culture medium (DMEM + 10% FBS + antibiotics). One-eighth (0.5 mL) of the released cells is redistributed to a new 100-mm culture tissue containing 8 mL fresh culture medium. The remainder (3.5 mL) is used for transfection.
5. Coverglass chips coated with polylysine are prepared prior to seeding with CHO cells to be transfected. Chips are prepared by soaking them in 100% ethanol, rinsing with water, and subsequently layering them with 2 mL 0.001% polylysine solution for 30 min (*see* **Note 8**). Chips are then rinsed of polylysine once

with water and then twice with PBS and allowed to air-dry in a tissue culture hood under ultraviolet light.

6. Subconfluent CHO cells to be transfected are prepared by layering polylysine-coated coverglass chips held in a 35-mm tissue culture dish filled with 2 mL fresh culture medium with 10–20 μL passed cells (from the 3.5 mL remaining after **step 4**). Each dish normally contains 12–16 chips that are 5 × 5 mm.

7. One day after seeding, cells on chips are transfected with 0.3 μg (*see* **Note 9**) plasmid cDNA encoding each of the three ENaC subunits plus 0.5 μg of the pEGFP-F plasmid encoding the EGFP marker used to identify transfected cells (*see* **Subheading 3.2., step 2**). The total amount of exogenous cDNA applied to each 35-mm dish containing cells seeded on chips then is 1.4 μg (0.3 + 0.3 + 0.3 + 0.5).

8. For transfection, cells are rinsed twice with PBS (supplemented with 2 m*M* CaCl$_2$) and exposed to transfection medium overnight. Two hours after addition of transfection medium to CHO cells, this medium is supplemented with 10 μ*M* amiloride, with transfected cells maintained in amiloride until used. Transfection medium is prepared by adding to 100 μL DMEM (with no FBS or antibiotics) the cDNAs of interest followed by 10 μL Polyfect reagent (*see* **Note 10**). This mixture is immediately vortexed for 10 s and then incubated at room temperature for 10–20 min. The volume of the transfection medium is next raised to 0.6 mL by addition of 0.5 mL DMEM supplemented with FBS and antibiotics, with all 0.6 mL added to the 35-mm dish containing freshly washed (with PBS) CHO cells on polylysine-treated chips in 2 mL DMEM supplemented with FBS and antibiotics (final volume is 2.6 mL).

3.2. Analysis of Recombinant ENaCs Using the Patch Clamp Method

1. Patch clamp analysis of ENaC heterologously expressed in CHO cells employs standard equipment usage (*see* **Note 11**).

2. A chip containing CHO cells transfected with ENaCs is removed from the tissue culture incubator and placed into a perfusion chamber affixed to the stage of an inverted microscope. The chip is rinsed of tissue culture medium and amiloride with constant perfusion of the extracellular bath solution.

3. Cells positively transfected are identified by EGFP emissions on proper excitation (*see* **Note 12**). An example of a transfected cell with EGFP emissions is shown in **Fig. 2**.

4. After compensating offsets, a cell-attached, high-resistance (>1 gΩ) seal is formed on a positive cell by lowering the patch pipet to the cell with the micromanipulator and applying gentle suction. Seal formation is assessed by monitoring pipet resistance with resistance going from approx 2–3 to approx 6–10 mΩ after touching the cell and from there more than 1 gΩ on applying suction. Pipet capacitance is then compensated. For whole-cell voltage clamp experiments, this seal is ruptured with additional gentle suction to provide access to the intracellular compartment with cell capacitance (~8 pF) and serial resistances compensated (*see* **Note 13**). On going whole cell, patch seal resistance will decrease from more than 1 to approx 100–300 mΩ (at a holding potential of 30 mV; *see* **Note**

bright field   eGFP emissions

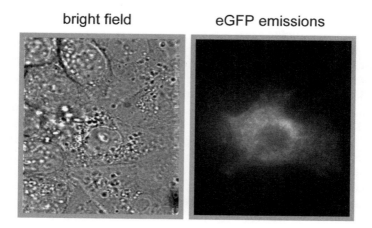

Fig. 2. Identifying a transfected cell. Chinese hamster ovary cells transfected with epithelial Na⁺ channels (ENaC) and EGFP-F where visualized in wide-field (left) and epifluorescence (EGFP) conditions (right) to identify a cell overexpressing the fluorescent reporter for positive transfection. We find good correlation between cells expressing the reporter and ENaC activity, with greater than 90% of the positive cells expressing EGFP also expressing the channel.

      14). For excised, outside-out patches, the pipet in the whole-cell configuration is slowly withdrawn from the cell, allowing the membrane to fold back upon the tip to form the outside-out seal (resistance should return to >1 gΩ; *see* **Note 15**).

5. After establishing the seal configuration of choice, patched membranes/cells are voltage clamped with inward currents referenced to ground downward and membrane potential equal to the pipet potential. In the whole-cell configuration, current through ENaCs is elicited by voltage ramping from 60 down to –100 mV over a 500-ms period. **Figure 3A** shows a representative voltage ramp protocol. Macroscopic currents are filtered at 1000 Hz, digitized at 2 kHz, and elicited/ recorded in episodic fashion using the clampex program to drive the patch clamp amplifier (*pClamp* software). Macroscopic current through ENaCs is identified by adding amiloride (10 µ*M*) to the extracellular bath solution. **Figure 3B** shows a typical overlay of macroscopic currents in a CHO cell expressing human ENaCs before (con.) and after (amil.) addition of amiloride. Shown in **Fig. 3C** is a complete experiment with currents elicited by a train of voltage ramps given every 2.5 s, with amiloride applied to the bath in the middle of the experiment (arrow notes 0 current level). An alternative to applying voltage ramps that is better suited for generating current–voltage relations and for investigating time-dependent channel events is to use voltage steps to elicit currents. **Figure 4A** shows representative macroscopic currents elicited by stepping from a holding potential of 30–100 mV and then down to –120 mV with steps of –20 mV for

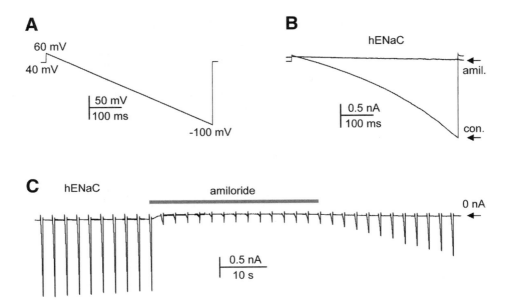

Fig. 3. Heterologous expressing and recording of human epithelial Na⁺ channels (ENaC) in Chinese hamster ovary (CHO) cells. (**A**) A typical voltage ramp used to elicit ENaC currents in transfected CHO cells. (**B**) A representative overlay of macroscopic currents in a voltage-clamped CHO cell expressing human ENaC before and after addition of amiloride to the bathing solution. Current was elicited with the voltage ramp shown in **A**. (**C**) Shown here is a typical series of ENaC currents in a voltage-clamped, transfected CHO cell elicited by a train of voltage ramps, such as that in **A**, before and after amiloride.

untransfected CHO cells (top) and cells overexpressing human (middle) and mouse (top) ENaCs before (left) and after (right) addition of amiloride to the extracellular bathing solution. ENaC activity is routinely reported as the amiloride-sensitive current density (current normalized to membrane area measured as capacitance) at −80 mV. Shown in **Fig. 4B** is a summary graph of the current density in untransfected CHO cells compared to the amiloride-sensitive current density in CHO cells expressing human and mouse ENaCs at −80 mV.

6. Current through individual ENaCs are recorded in excised, outside-out patches using clampex in the gap-free mode with current data filtered at 100 Hz and digitized at 400 Hz. For these experiments, membrane potential is held at 0 mV with amiloride added to the extracellular face of the channel in the bathing solution to confirm that the channel is indeed an ENaC (*see* **Note 16**). **Figure 5** shows a continuous current trace of a representative excised, outside-out patch containing at least four ENaCs before and after addition of amiloride to the extracellular face of the channel.

Staruschenko et al.

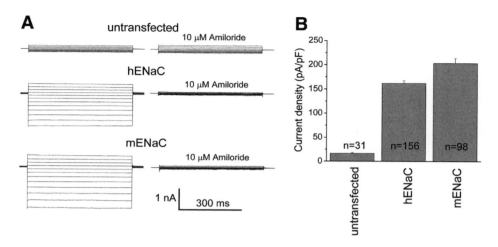

Fig. 4. Expression and recording of human and mouse epithelial Na$^+$ channels (ENaC) in Chinese hamster ovary (CHO) cells. (**A**) Macroscopic currents before (left) and after amiloride (right) in voltage-clamped CHO cells not expressing ENaC (top) and expressing human (middle) and mouse (bottom) ENaC. Currents were elicited by voltage stepping from the resting potential of 30–100 mV down to –120 mV by 20-mV steps. (**B**) Summary graph of macroscopic current density at –80 mV in voltage-clamped untransfected CHO cells and cells expressing human and mouse ENaC. Current for cells expressing human and mouse ENaC is the amiloride-sensitive current at this voltage. No amiloride-sensitive current is observed in untransfected cells.

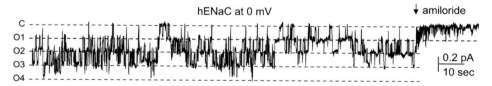

Fig. 5. Recording epithelial Na$^+$ channels (ENaC) in excised, outside-out patches made from Chinese hamster ovary (CHO) cells overexpressing the channel. Shown is a representative current trace from an excised, outside-out patch made from a CHO cell expressing ENaC. This patch was held at 0 mV and contains at least four ENaCs. Amiloride was added to (the bath) the extracellular face of these channels toward the end of the experiment.

4. Notes

1. cDNA may be purified using any number of methods. We strongly recommend that cDNA be high quality and isolated using a stringent maxi-/midiprep method or equivalent. We commonly use Wizard Plus Midiprep (Promega, A7640) to purify cDNA for transfection.

2. For proper expression of channel protein following transfection of CHO cells, it is required that the cDNA encoding the channel is positioned behind a mammalian promoter. We have had success driving expression from the cytomegalavirus and adenovirus major late promoters. It is likely that most mammalian promoters will sufficiently drive expression in CHO cells.

3. To facilitate identification of cells expressing ENaCs, we also often use a plasmid encoding a fusion protein of a channel subunit and a fluorophore (i.e., EGFP or EYFP; *see* **ref. *11***). This obviates the need to cotransfect pEGFP-F.

4. There are many different patch clamp amplifiers, analog-to-digital acquisition boards and programs, micromanipulators, isolation tables, microscopes, pipet pullers, perfusion chambers, and so on available for patch clamp analysis. Use those best suited to the experiments and personal preference.

5. We have listed our most commonly used intracellular pipet and extracellular bathing solutions. These can be adjusted to user preference and experimental conditions. By excluding K^+ from our solutions, we remove the possibility of contaminating K^+ currents when recording ENaC activity. With the whole-cell configuration, reagents, including small proteins, may be introduced through solution dialysis from the recording pipet into the intracellular milieu. Solution exchange is dependent on the size and shape of the patch pipet and the size of the molecule dialyzed (*see* **ref. *14***).

6. Many cell lines are suitable for heterologous expression of ion channel proteins and subsequent functional analysis of these channels using the patch clamp method. One should consider several factors when choosing a cell line for heterologous expression. Some of these are (1) whether the cell expresses a native channel similar to the one to be studied; (2) other background currents in the cell, which partly sets the noise of the system; (3) ease of patching the chosen cell; (4) signaling pathways in the cell; (5) the ability to express exogenous protein; (6) tolerance to transfection; (7) how the cell responds to prolonged culture and experimental conditions outside the incubator; and (8) tolerance to patch clamp solutions and reagents. We have successfully reconstituted and recorded ENaC currents in CHO, COS-7 (*6,7,10–13*), and Hek293 (unpublished observation) cells. Others have done so in NIH 3T3 fibroblasts (*15*) and MDCK cells (*16*).

7. There are several methods for continuing a stock cell line. We use this method because CHO cells tolerate trypsin well, with prolonged exposure having no observable adverse affects.

8. Polylysine is used to facilitate CHO cell adherence to coverglass. This sometimes can be particularly important for patch clamp experiments performed with constant bath perfusion. However, some types of cells suitable for heterologous

expression do not require plating on polylysine. In addition, some cells do not tolerate polylysine well; thus, in some instances, this coating may adversely affect experimental results. We have found that CHO and COS-7 cells tolerate polylysine coating well.

9. cDNA amounts can be titrated to increase or decrease channel expression levels (and current) as desired. Depending on the amount of ENaC activity desired, we use 0.2–0.5 µg cDNA for each subunit. ENaC activity is strongly correlated with the amount of cDNA used.

10. There are many means and reagents used to introducing cDNA into cells. We have had success with Polyfect, Superfect (Qiagen), and Lipofectemine Plus (Invitrogen, Carlsbad, CA) and using transfection, infection (*see* Chapter 2) and a Bialistic gene gun (*see* Chapter 3) as a mechanism to introduce exogenous cDNA into CHO cells.

11. The goal of this chapter is not to teach how to patch clamp but rather to provide a specific protocol for transfecting and recording ENaC currents in a mammalian expression system. For a detailed description of the patch clamp method, Sakmann and Neher (*1*) provided an excellent book.

12. We also strongly encourage the investigator to confirm channel expression with Western blot analysis or immunofluorescence. This becomes particularly important when the investigator believes cells are properly transfected but the channel of interest cannot be detected with electrophysiology (*see also* **Notes 2** and **3**).

13. Many patch clamp amplifiers can apply a brief but intense voltage step to facilitate seal rupture. On the Axopatch 200B, this is initiated with the "zap" button. With our conditions, we recommend not using this method for seal rupture.

14. An alternative to applying suction to rupture the seal is to use the perforated patch technique (*1,14*). We have successfully used 1.8 mg/mL amphotericin B (Calbiochem) in the pipet to form perforated, whole-cell patches (*see* **Fig. 1**). The benefit of this approach is that it does not disrupt the intracellular milieu and associated signaling pathways. It can, however, lead to problematic junction potentials.

15. It is also possible to form cell-attached and excised, inside-out patches with this preparation. Although the latter provides direct access to the intracellular face of the channel and the former retains cell signaling pathways, neither allows access to the extracellular face of the channel. When using either the cell-attached or excised, inside-out patch configuration, one must be aware that membrane potential is now negative pipet potential and adjust current orientation accordingly. In addition, bath and pipet solutions will be different compared to those described in the **Subheading 2**.

16. CHO cells have little endogenous channel activity. This is reflected by a resting membrane potential near 0 mV (*see* **Fig. 2**), virtually no background macroscopic currents (*see* **Fig. 4**), and few observable channels in excised patches made from untransfected cells. Indeed, with our solutions, the only native channel we have observed in excised patches is a small (<1 pS) cation channel that was rarely observed. In the whole-cell configuration, endogenous nonselective cation and Cl⁻ currents, though, have been activated on cell swelling and application of IGF-I.

References

1. Sakmann, B. and Neher, E. (1983) *Single-Channel Recording*, Plenum Press, New York, NY.
2. Aschroft, F .M. (2000) *Ion Channels and Disease*, Academic Press, London.
3. Garty, H. and Palmer, L. G. (1997) Epithelial sodium channels: function, structure, and regulation. *Physiol. Rev.* **77,** 359–396.
4. Snyder, P. M. (2002) The epithelial Na+ channel: cell surface insertion and retrieval in Na+ homeostasis and hypertension. *Endocr. Rev.* **23,** 258–275.
5. Hummler, E. and Horisberger, J. D. (1999) Genetic disorders of membrane transport. V. The epithelial sodium channel and its implication in human diseases. *Am. J. Physiol.* **276,** G567–G571.
6. Tong, Q., Gamper, N., Medina, J. L., Shapiro, M. S., and Stockand, J. D. (2004) Direct activation of the epithelial Na(+) channel by phosphatidylinositol 3,4,5-trisphosphate and phosphatidylinositol 3,4-bisphosphate produced by phosphoinositide 3-OH kinase. *J. Biol. Chem.* **279,** 22,654–22,663.
7. Booth, R. E., Tong, Q., Medina, J., Snyder, P. M., Patel, P., and Stockand, J. D. (2003) A region directly following the second transmembrane domain in gamma ENaC is required for normal channel gating. *J. Biol. Chem.* **278,** 41,367–41,379.
8. McDonald, F. J., Price, M. P., Snyder, P. M., and Welsh, M. J. (1995) Cloning and expression of the B- and y-subunits of the human epithelial sodium channel. *Am. J. Physiol.* **268,** C1157–C1163.
9. Puck, T. T., Cieciura, S. J., and Robinson, A. (1958) Genetics of somatic mammalian cells. III. Long-term cultivation of euploid cells from human and animal subjects. *J. Exp. Med.* **108,** 945–956.
10. Staruschenko, A., Nichols, A., Medina, J. L., Camacho, P., Zheleznova, N. N., and Stockand, J. D. (2004) Rho small GTPases activate the epithelial Na+ channel. *J. Biol. Chem.* **279,** 49,989–49,994.
11. Staruschenko, A., Medina, J. L., Patel, P., Shapiro, M. S., Booth, R. E., and Stockand, J. D. (2004) Fluorescence resonance energy transfer analysis of subunit stoichiometry of the epithelial Na+ channel. *J. Biol. Chem.* **279,** 27,729–27,734.
12. Tong, Q. and Stockand, J. D. (2005) Receptor tyrosine kinases mediate epithelial Na+ channel inhibition by epidermal growth factor. *Am. J. Physiol.* **288,** F150–F161.
13. Staruschenko, A., Patel, P., Tong, Q., Medina, J. L., and Stockand, J. D. (2004) Ras activates the epithelial Na(+) channel through phosphoinositide 3-OH kinase signaling. *J. Biol. Chem.* **279,** 37,771–37,778.
14. Hamill, O. P., Marty, A., Neher, E., Sakmann, B., and Sigworth, F. J. (1981) Improved patch-clamp techniques for high-resolution current recording from cells and cell-free membrane patches. *Pflugers Arch.* **391,** 85–100.
15. Gilmore, E. S., Stutts, M. J., and Milgram, S. L. (2001) SRC family kinases mediate epithelial Na+ channel inhibition by endothelin. *J. Biol. Chem.* **276,** 42,610–42,617.
16. Ishikawa, T., Marunaka, Y., and Rotin, D. (1998) Electrophysiological characterization of the rat epithelial Na+ channel (rENaC) expressed in MDCK cells. Effects of Na+ and Ca2+. *J. Gen. Physiol.* **111,** 825–846.

2

Overexpression of Proteins in Neurons Using Replication-Deficient Virus

Richard M. Ahlquist and Jane M. Sullivan

Summary

Overexpression of proteins is a powerful way to determine their function. Until recently, the low efficiency of neuronal transfection has made it difficult to use overexpression and structure–function studies to investigate the role of neuronal proteins in their native environment. The development of neurotrophic viral systems has overcome the obstacle of low efficiency and allows for unprecedented opportunities to use biochemical and electrophysiological techniques to assess the effects of overexpressing wild-type or mutant proteins in neurons. Here, a general protocol for the production of replication-deficient Semliki Forest virus constructs directing the overexpression of proteins of interest in cultured mammalian neurons is described.

Key Words: Cultured neurons; infection; overexpression; replication-deficient virus; Semliki Forest virus; structure/function; virions.

1. Introduction

Structure–function studies provide valuable information about the mechanisms underlying protein function. Most structure–function studies to date have used heterologous expression systems that place constraints on the proteins that can be studied and the questions that can be addressed. This has been a particular problem for the study of neuronal protein function given the highly specialized nature of the neuron itself. One of the greatest concerns with the use of heterologous expression systems for overexpression and structure–function studies of neuronal proteins is the possibility that the protein will not be properly processed or trafficked outside its native environment. In addition, proteins that may ordinarily associate with the protein of interest (POI) may not be expressed in the heterologous system. To get around these problems, we

From: *Methods in Molecular Biology, vol. 337: Ion Channels: Methods and Protocols*
Edited by: J. D. Stockand and M. S. Shapiro © Humana Press Inc., Totowa, NJ

overexpress wild-type and mutant versions of neuronal proteins with functions we are interested in studying in cultured mammalian central nervous system neurons using replication-deficient virus.

Here, we describe the generation of replication-deficient Semliki Forest virus virions. The first step is the polymerase chain reaction (PCR)-based subcloning of the nucleotide sequence encoding the POI into a mammalian expression vector, pIRES2-EGFP (enhanced green fluorescent protein). The pIRES2-EGFP vector directs the production of the POI separately from a reporter protein (EGFP) through an intervening internal ribosomal entry site (IRES) sequence. The first subcloning is followed by a second subcloning of the POI-IRES-EGFP cassette into the replication-deficient Semliki Forest virus vector (pSFV). Virions are produced after electroporation of RNA derived from the POI-IRES-EGFPpSFV1 construct, along with helper RNA, into baby hamster kidney (BHK) cells. Virions are activated and used to infect cultured neurons. Finally, immunocytochemistry is performed to verify expression of the POI.

2. Materials

2.1. PCR Amplification of POI Complementary DNA

1. Molecular biology grade (MBG) water (Eppendorf, Westburg, NY).
2. Thermostable proofreading DNA polymerase: Vent polymerase and buffer (New England Biolabs, Ipswich, MA).
3. Dimethyl sulfoxide (DMSO).
4. dNTPs: mix of 10 mM each, from 100 mM deoxynucleotide 5'-triphosphate (dNTP) set, PCR grade (Invitrogen, Carlsbad, CA) in MBG water.
5. Project-specific primers (Integrated DNA Technologies, Coralville, IA).
6. Template DNA encoding POI.
7. Mineral oil.
8. 1X Tris-borate ethylenediaminetetraacetic acid (EDTA) (TBE) buffer: 89 mM Tris-borate, 2 mM EDTA (from 10X TBE; Fisher, Pittsburgh, PA).
9. 10 mg/mL ethidium bromide stock (Sigma, St. Louis, MO; use caution as this is toxic).
10. 1% agarose (Invitrogen) gel in 1X TBE with 4 μL ethidium bromide stock/100 mL.
11. 10X dye loading buffer: 15% Ficoll-400, 0.25% bromophenol blue, 0.25% xylene cyanol FF in MBG water.
12. 5X PCR loading buffer: 1:1 mixture of 10X dye loading buffer and 10X TBE.
13. 1 Kb DNA ladder (Invitrogen): run 0.5 μg/lane using 5X PCR loading buffer.
14. QIAquick PCR Purification Kit (Qiagen, Germantown, MD).

2.2. Subcloning of PCR-Generated POI Complementary DNA Into the pIRES2-EGFP

1. pIRES2-EGFP vector (Clontech, Mountainview, CA).
2. Project-specific restriction enzymes (REs; New England Biolabs, Roche, Indianapolis, IN).

3. 10X bovine serum albumin (BSA): make from 100X BSA (10 mg/mL; New England Biolabs) in MBG water.
4. Calf intestine alkaline phosphatase (CIAP; MBI Fermentas, Burlington, Ontario, Canada).
5. 1X Tris-acetate EDTA (TAE) buffer: 40 mM Tris-acetate, 1 mM EDTA (from 10X TAE; Fisher).
6. 1% low-melting-point agarose (LMPA; Invitrogen) gel in 1X TAE with 4 µL ethidium bromide stock/100 mL.
7. Phenol/chloroform/isoamyl alcohol 25:24:1 (Sigma Fluka, St. Louis, MO; use caution as this mix is caustic).
8. Chloroform/isoamyl alcohol 49:1 (Sigma Fluka).
9. Pellet Paint coprecipitant (Novagen, Madison, WI).
10. Millipore UltraFree FilterSpin columns (Fisher).
11. Rapid DNA Ligation Kit (Roche).
12. One Shot competent *Escherichia coli* (Invitrogen).
13. Luria-Bertani (LB) medium.
14. Bacto Agar for plates (BD Biosciences, Clontech).
15. LB with kanamycin (KAN; 50 µg/mL).
16. KAN plates: LB agar plates with kanamycin (50 µg/mL).
17. QIAprep Spin Miniprep Kit (Qiagen).

2.3. Subcloning of Complementary DNA for POI-IRES-EGFP Into pSFV

1. REs and other materials as in **Subheading 2.2.**
2. pSFV and pSFVHelper2 vectors (Invitrogen; *see* **Note 1**).
3. Guanosine 5'-triphosphate (GTP)/cytidine 5'-triphosphate (CTP) mix: Mix of 1 mM each from PCR grade 100 mM dNTP set (Invitrogen) in MBG water.
4. Large (Klenow) fragment DNA polymerase I (Klenow; New England Biolabs).
5. LB with ampicillin (AMP; 150 µg/mL).
6. AMP plates: LB agar plates with ampicillin (150 µg/mL).

2.4. Linearization of Template DNA

1. Restriction enzyme: *Spe*I preferred (New England Biolabs, Roche).
2. Phenol/chloroform/isoamyl alcohol 25:24:1 (Sigma Fluka).
3. Chloroform/isoamyl alcohol 49:1 (Sigma Fluka).
4. Pellet Paint coprecipitant (Novagen).

2.5. RNA In Vitro Transcription

1. SP6 mMessage mMachine Kit (Ambion, Austin, TX).
2. 1X MOPS buffer: from 10X MOPS buffer (Fisher) in MBG water.
3. Denaturing formaldehyde gel: Place 0.5 g agarose in 35 mL MBG H$_2$O, heat in microwave, and cool to approx 60°C, and add 10 mL formaldehyde (*do not microwave formaldehyde*) and 5 mL 10X MOPS.
4. RNA loading buffer: 72 µL formamide, 26 µL formaldehyde, 16 µL 10X MOPS buffer, 8 µL glycerol, 18 µL MBG H$_2$O, and 2 µL ethidium bromide stock (10 mg/mL).
5. 10X dye loading buffer (*see* **Subheading 2.1.**, **item 10**).

2.6. Virion Production and Activation

1. BHK cells (ATCC, Manassas, VA).
2. Dulbecco's modified Eagle's medium (DMEM; Gibco/Invitrogen).
3. BHK medium: 95 mL DMEM, 5 mL fetal calf serum (Gibco), 0.25 mL pen/strep antibiotic (Gibco/Invigrogen).
4. Trypsin/EDTA (0.25% trypsin/1 mM EDTA; Invitrogen).
5. 1X Ribonuclease (RNase)-free phosphate-buffered saline (PBS): from 10X PBS (Fisher) in MBG water.
6. Chymotrypsin (Worthington Biochem, Lakewood, NJ): 2 mg/mL in PBS with 0.9 mM CaCl$_2$ and 0.5 mM MgCl$_2$.
7. Aprotinin (Roche): 6 mg/mL in PBS with 0.9 mM CaCl$_2$ and 0.5 mM MgCl$_2$.

2.7. Infection of Neuronal Culture

1. Cultured neurons.
2. Activated virions.

2.8. Immunocytochemical Confirmation of Protein Expression

1. Fix solution: 4% paraformaldehyde plus 4% sucrose in PBS.
2. Permeabilization buffer: 2% Triton X-100 in PBS.
3. Blocking buffer: 5% BSA (Sigma) in PBS.
4. 1° antibody solution: approx 1 μL antibody/mL in blocking buffer.
5. 2° antibody solution: approx 2 μL antibody/mL in blocking buffer.

3. Methods

Be aware that most institutions require permission to work with replication-deficient viruses. For replication-deficient semlike forest virus, the National Institutes of Health recommends biosafety level 3 practices in a biosafety level 2 environment, including use of a type II/class A biosafety cabinet.

3.1. PCR Amplification of POI Complementary DNA

It is almost always necessary to use PCR to introduce the three RE sites required for the two sequential subcloning steps employed for generation of the viral construct (*see* **Fig. 1**).

1. Primer design: select three RE sites to add to either side of the POI. Complementary DNA (cDNA) sequence in the PCR product: two at the 5' end, upstream of a Kozak consensus sequence and the Start codon, and one at the 3' end, immediately following the Stop codon. RE sites 1 and 2 are used to subclone the PCR product, and RE site 3 is used for the subsequent subcloning of the POI-IRES-EGFP cassette into the pSFV viral vector (*see* **Subheadings 3.2.** and **3.3.**). A Kozak sequence (GCCACC) is introduced just before the Start codon to ensure robust expression of the POI. The 5' primer encodes RE site 1, RE site 3, the Kozak sequence, the Start codon, and 15–20 additional bases perfectly matched to the POI cDNA sequence. The 3' primer

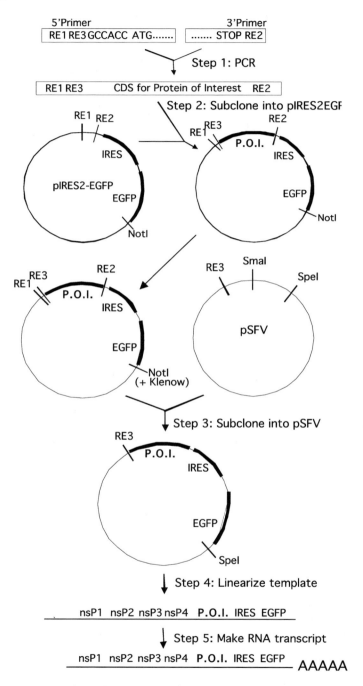

Fig. 1. Flowchart of **Subheadings 3.1–3.5.**

encodes 15–20 bases perfectly matched to the POI cDNA sequence, the Stop codon, and RE site 2. RE sites 1–3 must not be present in the cDNA encoding the POI. Be careful to note whether the new sequence created by the primers introduces any additional restriction sites. A GC lock at the 3' end of each primer is recommended.

In choosing RE site 3, note that the multiple cloning site (MCS) for pSFV contains just three restrictions sites: *Bam*HI, *Sma*I (a blunt cutter), and *Xma*I (an isoschizomer of *Sma*I that leaves an overhang). We now use a homemade pSFV variant with an improved MCS; we also use a homemade pIRES2-EGFP variant in which the *Bam*HI at the end of the MCS site has been removed, allowing us to use *Bam*HI as RE3.

Keep the initial melting temperature of your primers 65°C or above. Melting temperature is determined by the number and composition of nucleotides in the primers. There are many ways to estimate the melting temperature, but a quick and easy way is to multiply the number of A or T bases by 2, multiply the number of G or C bases by 4, then add the two numbers (this method will overestimate the melting temperature for long sequences, but it usually does not matter). To calculate the initial melting temperature, use only the perfectly matched bases between the primers and the POI template DNA sequence (i.e., do *not* include the bases encoding the RE sites or the Kozak sequence unless already present). After the first few cycles of PCR, sufficient PCR product builds up to serve as template itself; the primers have a much higher melting temperature with this PCR-derived template because of the additional bases of perfectly matched sequence that are not included in the initial melting temperature calculation.

2. PCR: to a PCR tube, add 34.5 µL MBG H_2O, 5 µL 10X Vent polymerase buffer, 2.5 µL DMSO, 1 µL dNTPs (10 m*M* each), 1 µL template DNA encoding POI, and 1 µL Vent polymerase for a total reaction volume of 45 µL. Overlay with 50 µL mineral oil. Place tube in PCR machine and start program (*see* below). After the PCR machine reaches 80°C, start the reaction by adding 2.5 µL of each primer (20 µ*M* stock in MBG water), for a final reaction volume of 50 µL. This "hot start" method increases the specificity and yield of the PCR product, as does the addition of DMSO.

The following is an example PCR program for a 2.3-kb POI cDNA:

> 1 cycle:
> - 95°C for 1.5 min (melting step)
> - 60°C for 2 min (annealing step)
> - 72°C for 2 min 20 s (extension step)
>
> 2 cycles:
> - 95°C for 30 s
> - 60°C for 2 min
> - 72°C for 2 min 20 s
>
> 22 cycles:
> - 95°C for 30 s
> - 65°C for 1 min
> - 72°C for 2 min 20 s
>
> Hold at 4°C

The annealing step temperature is raised (and the duration reduced) after the first three cycles because sufficient PCR product, perfectly matched to the entire length of the 5' and 3' primers, has been generated to serve as a template for subsequent rounds. The duration of the extension cycle is calculated assuming a polymerization rate for Vent polymerase of 1000 bases/min.

3. Diagnostic gel: verify the quality and yield of the PCR product (4 µL PCR product + 1 µL 5X PCR loading buffer) on a 1% agarose TBE gel in TBE running buffer using 1-kb DNA ladder. 5X PCR loading buffer ensures that low-molarity PCR samples are retained in the gel wells.

4. Purification of the PCR product: purify the PCR product with Qiagen's QIAquick PCR Purification Kit following the manufacturer's instructions. Purification is necessary to remove the proofreading Vent polymerase, which will remove overhangs created during the restriction digest described in the **Subheading 3.2.**

3.2. Subcloning of PCR-Generated POI cDNA Into the pIRES2-EGFP

1. Preparation of the PCR insert and vector: both the PCR product and the pIRES2-EGFP vector are cut with REs 1 and 2. Cut 14 µL purified PCR product for the insert (less if yield is very high). Cut 2 µg pIRES2-EGFP vector DNA brought up to a total of 14 µL volume with MBG water. After digestion with the REs, the vector is treated with phosphatase to reduce background (vector religating without insert). To a final volume of 20 µL, add 14 µL PCR DNA (or 2 µg pIRES2-EGFP in 14 µL MBG H_2O), 2 µL 10X restriction buffer, 2 µL BSA (if required), 1 µL RE 1, and 1 µL RE 2. Incubate restriction digest at 37°C for 2 h. Include BSA if either enzyme calls for it. After 2 h, add 0.5 µL CIAP to vector only (do *not* add CIAP to PCR DNA digest) and incubate for an additional 0.5 h at 37°C.

2. LMPA gel purification of PCR insert: add 2 µL 10X dye loading buffer to the restriction digest and run the cut PCR product out on a 1% LMPA gel in TAE running buffer. Cut DNA band out of the gel (using an adjustable ultraviolet illuminator, if possible, to minimize exposure of the DNA to ultraviolet light), mince the band, and place in a Millipore FilterSpin column; spin at maximum speed for 15 min.

3. Organic extraction of vector: add 80 µL MBG H_2O to digested pIRES2-EGFP sample to bring the volume up to 100 µL. Add an equal volume (100 µL) phenol/chloroform/isoamyl alcohol (caution: caustic). Vortex for 1 min. Spin for 3 min at maximum speed in a microcentrifuge. Transfer 100 µL of the aqueous (upper) phase to a fresh microcentrifuge tube. Add 100 µL of chloroform/isoamyl alcohol, vortex 1 min, spin 3 min, and transfer 90 µL of the aqueous phase to a fresh microcentrifuge tube.

4. Precipitate vector: follow manufacturer's instructions for Novagen's Pellet Paint and resuspend pellet in 40 µL MBG H_2O. Pellet Paint allows for a quick and efficient precipitation.

5. Ligation of PCR insert into pIRES2-EGFP vector: ligate the gel-purified cut PCR DNA into the prepared pIRES2-EGFP vector following manufacturer's instructions for the Roche Rapid DNA Ligation Kit. Briefly, add to make a final reaction volume

of 21 µL, 1 µL digested and purified pIRES2-EGFP, 7 µL digested and purified insert DNA for POI, 2 µL solution 2, 10 µL solution 1, and 1 µL solution 3 (enzyme). Incubate at room temperature for 15 min. For a vector-only control (to test the background occurrence of vector closing without an insert by performing the ligation reaction without any insert DNA added), put all of the above in a separate tube but replace PCR insert DNA with 7 µL MBG H$_2$O.

6. Transformation of competent *E. coli*: follow manufacturer's instructions for transformation of One Shot competent cells. Plate all of transformation mix on KAN plates and place in 37°C incubator overnight.

7. DNA miniprep: start four to six miniprep cultures with LB with KAN and put tubes in 37°C shaking incubator overnight (14–18 h). Purify plasmid DNA using QIAprep Spin Miniprep Kit according to manufacturer's instructions.

8. Diagnostic restriction digest and gel electrophoresis: cut the prepped DNA with REs 1 and 2 for 1 h at 37°C in a 10 µL reaction containing 4 µL DNA, 3 µL H$_2$O, 1 µL 10X restriction buffer, 1 µL BSA (as needed), and 0.5 µL REs 1 and 2. At end of digest, add 1 µL 10X dye loading buffer to sample and run cut product out on 1% agarose TBE gel (*see* **Subheading 3.1.**, **step 3**).

9. Sequence plasmid: sequence one (or more) of the plasmids having the correct restriction digest band pattern to verify that no base changes have been introduced by PCR (*see* **Note 2**).

3.3. Subcloning cDNA for POI-IRES-EGFP Into pSFV

The cDNA encoding the POI and EGFP, and the intervening IRES sequence, is excised from pIRES2-EGFP, purified, and inserted into the pSFV vector.

1. Restriction digest and Klenow treatment of POI-IRES-EGFP insert: in a reaction with a final volume of 27 µL, mix POIpIE (2 µg) in 20 µL MBG H$_2$O, 3 µL 10X restriction buffer, 3 µL 10X BSA, and 1 µL *Not*I (*see* **Note 3**). Cut for 1 h at 37°C. Add 1 µL 1 m*M* GTP/CTP mix (*Not*I site is all Gs and Cs) and 1 µL Klenow enzyme. Incubate for 30 min at room temperature to fill in *Not*I-generated overhang and create a *Sma*I-compatible blunt end (*see* **step 2**). After 30 min, inactivate Klenow by incubating at 75°C for 10 min. After cooling to 37°C, add 1 µL RE 3 and cut for 1 h at 37°C (*see* **Note 4**).

2. Restriction digest and phosphatase treatment of pSFV vector: to a final volume of 20 µL, add 14 µL pSFV (1 µg) in MBG H$_2$O, 2 µL 10X restriction buffer, 2 µL 10X BSA (if required), 1 µL RE 3, and 1 µL *Sma*I. Incubate for 2 h at 37°C, then add 0.5 µL CIAP and incubate for an additional 0.5 h at 37°C.

3. LMPA purification of insert (*see* **Subheading 3.2.**, **step 2**): use 3 µL 10X dye loading buffer.

4. Organic extraction and Pellet Paint precipitation of pSFV vector (*see* **Subheadings 3.2.**, **step 3** and **3.2.**, **step 4**).

5. Ligation of insert into vector (*see* **Subheading 3.2.5.**, **step 5**): use 7 µL POI-IRES-EGFP insert.

6. Transformation of competent *E. coli*: follow manufacturer's instructions for transformation of One Shot competent cells. Plate all 300 μL of transformation mix on AMP plate and place in 37°C incubator overnight.
7. DNA miniprep: Start four to six miniprep cultures with LB with AMP and put tubes in 37°C shaking incubator overnight (14–18 h). Purify plasmid DNA using QIAprep Spin Miniprep Kit according to manufacturer's instructions.
8. Diagnostic restriction digest and gel electrophoresis: cut with RE 3 and *Spe*I (*see* **Subheading 3.4.**, **step 1**) to identify correct POI-IRES-EGFPpSFV constructs.

3.4. Linearization of Template DNA

Linearized template DNA is used for in vitro transcription of RNA that is electroporated into BHK cells to make the virions. Prepare template DNA of your construct and pSFVHelper2 by cutting plasmid DNAs with *Spe*I (*see* **Note 5**).

1. Linearize DNA in a final volume of 40 μL by adding 13 μg DNA in 30 μL MBG H$_2$O, 4 μL10X restriction buffer, 4 μL 10X BSA, 2 μL *Spe*I. After 2 h, add 60 μL MBG H$_2$O to bring total volume to 100 μL.
2. Organic extraction: *see* **Subheading 3.2.**, **step 3**.
3. Precipitate linearized DNA with Pellet Paint: *see* **Subheading 3.2.**, **step 4** Resuspend in 15 μL MBG H$_2$O.

3.5. RNA In Vitro Transcription

Use gloves and RNase-free tubes during this and subsequent steps to prevent RNA degradation by the RNases that are ubiquitously present on skin.

1. In vitro RNA transcription from template DNA: follow manufacturer's instructions for SP6 mMessage mMachine. To a final reaction volume of 20 μL, add 10 μL 2X NTP/CAP solution, 2 μL 10X reaction buffer, 4 μL linear template DNA, 2 μL GTP (required for long transcripts), and 2 μL SP6 enzyme mix. Incubate for 2 h at 37°C, store at −20°C (or colder).
2. Denaturing gel electrophoresis assessment of RNA quality and quantity: remove 1 μL RNA reaction mix and add to 3 μL RNA dye loading buffer; run out on a denaturing formaldehyde gel to test for quality and quantity. Load one lane with 1 μL 10X loading buffer to monitor progress through the gel. Good RNA will run as a tight, bright band (sometimes as a doublet).

3.6. Virion Production and Activation

POI-IRES-EGFPpSFV RNA and SFVHelper2 RNA are electroporated into BHK cells to produce *inactive* replication-deficient virions encoding the POI and EGFP. A safety feature of the SFV system is that chymotrypsin treatment is required to activate the virions before they are able to infect neurons. Check with your local safety administrators for the requirements your institution may have for working with the replication-deficient Semliki Forest virus.

1. Grow and harvest BHK cells: grow BHK cells in BHK medium at 37°C in 175-cm² tissue culture flasks until 80–100% confluent. One 175-cm² flask will yield approx 0.5–2 × 10⁷ cells at 80–100% confluence (~1 × 10⁷ cells are required for each batch of virions). Remove BHK medium and briefly rinse flask bottom with approx 2–5 mL DMEM (or other serum-free solution). Add 5–7 mL trypsin/EDTA to the flask and incubate 5 min to allow cells to lift from the flask bottom. Firmly tap the flask on the side to completely free the cells. Add 5 mL DMEM, rinsing the surfaces of the flask, and transfer the contents of the flask to a 15-mL conical vial. Spin 4 min in a clinical centrifuge to pellet cells.

   Carefully pour off the supernatant from the previous spin. Add 5 mL DMEM and triturate until cells are fully resuspended (taking care to avoid air bubbles). Add another 5 mL DMEM for a total volume of 10 mL. Spin 4 min in the clinical centrifuge to pellet cells. Carefully pour off the supernatant. Resuspend all cells in 10 mL RNase-free PBS (combining cells from multiple flasks). Place a drop of the cell suspension on a hemocytometer and count cells. Spin remaining suspension for 4 min in a clinical centrifuge to pellet cells, pour off supernatant, and resuspend cells in an appropriate volume of RNase-free PBS to give a concentration of approx 1 × 10⁷ cells/mL.

2. Electroporation of RNA into BHK cells: add 0.8 mL BHK cell suspension to a cuvet and then add 9 μL POI-IRES-EGFP RNA and 9 μL SFVHelper2 RNA. Place cuvet on ice.

   This procedure assumes use of a Bio-Rad Gene Pulser II and Capacitance Extender Plus. Electroporate immediately after adding the RNA with Gene Pulser set at 0.4 kV and 900 μF. The time constant should read 12–15 ms. Place the cuvet on ice for 5 min. To a 60-mm tissue culture dish, add 5 mL BHK medium and the contents of the cuvet, avoiding the mucilaginous debris produced during electroporation that floats at the surface (*see* **Note 6**).

3. Production of virions: incubate transfected BHK cells for 48 h at 31°C and harvest the medium (which contains the released virions). Growing the cells at 31°C after electroporation increases the titer of the virion stocks. Freeze virion stocks overnight at –20°C, then store at –80°C.

4. Activation and storage of virions: to 0.5 mL virion stock, add 50 μL chymotrypsin and incubate for 40 min at room temperature. After 40 min, add 55 μL aprotinin and incubate for 5 min at room temperature. Store activated virion stock at –20°C for up to a month (*see* **Note 7**).

3.7. Infection of Neuronal Culture

Add 5–50 μL activated virion stock per milliliter neuronal culture medium and incubate for 4–48 h. Infection efficiency for virion stocks varies widely from virion prep to virion prep, and the appropriate volume for each batch must be determined empirically. In addition, different neuronal culture preparations can be more or less amenable to infection; this also can only be determined empirically. It takes about 10 h for the EGFP to become clearly detectable in infected neurons, and cell health often declines 24–48 h after infection. Adding more virion stock will increase the number of infected neurons (and start to infect astrocytes), but cell health is often compromised; this may be acceptable for biochemical assays for which maximal infection efficiency is required, and harvesting of neurons can take place earlier than 10 h (significant amounts of protein are made even in the first few hours after addition of the virion stock, although they may not have a chance to be trafficked to their proper destination).

3.8. Immunocytochemical Confirmation of Protein Expression

Immunocytochemistry is used to confirm expression of the POI. Once protein expression has been confirmed for a particular POI-IRES-EGFPpSFV construct, GFP fluorescence is sufficient to indicate the presence of POI.

1. Fix cells: replace culture medium with appropriate volume of fix solution and incubate at room temperature for 20 min. Replace solution with permeabilization buffer and incubate at 4°C for 10 min. Replace solution with blocking buffer and incubate at 4°C for 1 h.
2. Label cells with primary antibody: replace blocking buffer with an appropriate volume (just enough to cover is usually sufficient) of 1° antibody solution for 1–12 h at 4°C. Remove 1° antibody solution and wash three times with blocking buffer for 5 min at room temperature, rocking slowly.
3. Labeling cells with secondary antibody: replace blocking buffer with 2° antibody solution and rock slowly in the dark (or cover with foil) for 1 h at room temperature. Wash three times with PBS, rocking slowly in the dark for 10 min at room temperature.
4. Mount cover slips on slides and visualize using appropriate detection technique (e.g., fluorescent microscopy).

4. Notes

1. Invitrogen has discontinued the SFV Gene Expression System.
2. So-called silent mutations, in which a base change does not alter the amino acid sequence, are acceptable as long as the mutation does not introduce any new unwanted restriction sites.
3. If there is a *Not*I site in the coding sequence of your POI, then modify the *Not*I-plus-Klenow strategy using the next unique downstream restriction site in pIRES2-EGFP; note that the *Xba*I site just beyond the *Not*I site is methylated and will not cut unless DNA is grown in a dam⁻ host.
4. Depending on the size of the sequence encoding your POI, it may be difficult to distinguish your "insert" band from the (unwanted) "left-over pIE vector" band when you run the cut plasmid out on an LMPA gel (**Subheading 3.3., step 3**) if the POI-pIRES2-EGFP is cut with only two enzymes. For these cases, when you add RE 3 to the restriction digest mix, include an additional enzyme that cuts the vector at an appropriate location to allow ample separation of bands.

5. *Spe*I is the manufacturer's recommendation for the linearizing enzyme. If you have an *Spe*I site within the coding sequence for your POI, then you must choose an alternate unique cutter downstream of *Spe*I. Although the SFV Gene Expression System manual suggests using *Sap*I, we have found that this enzyme cuts at locations other than its predicted restriction sites; other enzymes to consider are *Pvu*I, *Xmn*I, and *Sph*I. We have successfully generated virions using template linearized with *Sph*I, the site farthest from the *Spe*I site.

6. For a quick and crude assessment of transfection efficiency, add a sterile 12-mm cover slip to the tissue culture dishes; this cover slip can be removed 24–48 h after electroporation and inspected under fluorescence to check for production of GFP. In our best preps, at least 70% of the BHKs are green.

7. Virion stocks lose their infection efficiency with time, even when stored at −80°C without thawing. Some virion stocks go bad after a few months; others remain usable for up to a year. Repeated freezing and thawing reduces infection efficiency, so small working aliquots are recommended once approximate infection efficiencies have been empirically determined.

3

Exogenous Expression of Proteins in Neurons Using the Biolistic Particle Delivery System

Nikita Gamper and Mark S. Shapiro

Summary

Exogenous expression of genes in mammalian neurons represents a substantial experimental challenge because of the low efficiency of commercially available liposomal transfection reagents for nondividing cells and considerable toxicity of viral transfection systems. In this chapter, we discuss application of the "biolistic" particle delivery system for heterologous expression of genes in primary neuron cultures. The method is based on the direct introduction of cDNA of interest into the nucleus by penetration with DNA-coated gold particles. With this approach, cDNA expression is independent of cell cycling and proliferation and is similar to intranuclear microinjection, with both avoiding cDNA delivery through the cytosol. Examples of successful transfection using PDS of rat superior cervical ganglion and trigeminal ganglion neurons are discussed.

Key Words: Gene gun; neuron; particle delivery; PI(4,5)P2; transfection.

1. Introduction

The advent of molecular cloning has made it possible to express exogenous genes in a variety of cells to probe the molecular mechanisms of channel physiology and regulation. For most cell lines, there are a number of commercially available transfection reagents that make such expression easy and reliable. These reagents typically facilitate the transport of DNA plasmids across the plasma membrane into the cytoplasm, from where mitotic mechanisms incorporate the DNA into the nucleus. However, these reagents often fail in the case of postmitotic (nondividing) cells such as neurons. Thus, other methods are often required for exogenous expression in those cells.

We here describe the use of the biolistic particle delivery system (PDS) for the cases of superior cervical ganglion (SCG) sympathetic neurons, and sensory

From: *Methods in Molecular Biology, vol. 337: Ion Channels: Methods and Protocols*
Edited by: J. D. Stockand and M. S. Shapiro © Humana Press Inc., Totowa, NJ

neurons of the trigeminal ganglion (TG). The former are noradrenergic neurons that receive cholinergic inputs from the central nervous system and have served as a model neuron for many studies of channel modulation, especially voltage-gated Ca^{2+} and K^+ channels *(1)*. The latter are a heterogeneous population of neurons that sense pain, touch, and temperature and innervate the facial region, including the mouth. We have published concerning use of the PDS on both these neuronal types *(2–5)*. Our successful use of the PDS in SCG neurons is predicated on the previous work of the Nerbonne lab, which performed a number of elegant studies on the molecular heterogeneity and physiological role of Kv K^+ channels in SCG cells *(6,7)*.

The PDS, which allows the use of the same cDNA vectors commonly used to transiently transfect cell lines, is in our opinion less complicated compared to viral expression systems, which require subcloning of genes into viral vectors and further viral packaging. For PDS, gold microspheres approx 1 μm in diameter are coated with the cDNA of interest to form microcarriers. Multiple plasmids can be used at the same time. The coated microspheres are then projected at high velocity at cells cultured in a dish, utilizing the gas pressure of an ordinary helium gas cylinder. The gold particles penetrate the cell but not the plastic bottom of the dish.

Any cells that happen to have a microsphere land in the nucleus have a high probability of expressing the gene coded for by the cDNA. Thus, the probability of a given cell expressing the protein coded for by the transfected cDNA is dependent on the scatter of the microspheres on bombardment. A success rate of 5–10% is maximal, which, although too low for biochemical analyses, is satisfactory for electrophysiological assays, especially if the cDNAs of interest are introduced with a marker, such as green fluorescent protein (GFP).

In this chapter, we show results using the PDS on SCG and TG cells. We do not describe here methods for isolating, dissociating, and culturing SCG or TG neurons or for cloning and isolating plasmid cDNA, but rather describe the method for introducing exogenous cDNA in nondividing cells such as neurons. The advantages of the PDS system include the need for little pre- and postbombardment manipulation of the cells and cDNA plasmid and the ability to transform a variety of cell types.

2. Materials

2.1. Equipment

1. We use the Biolistic PDS-1000/He Particle Delivery System (Bio-Rad, Hercules, CA), which is the biolistic equipment of choice for cultured cells. The Helios gene gun sold by the same company is preferable for DNA delivery into tissue and small organisms but is less suitable for cultured cells and thus is not discussed here.

2. A vacuum source capable of 15 mmHg vacuum is required. A typical laboratory vacuum source is sufficient.
3. A standard laboratory high-pressure, high-purity helium tank (2400–2600 psi) and associated regulator are required to optimize bombardment conditions.
4. Standard vortex mixer with platform attachment.
5. The 500 optimization kit (Bio-Rad) provides the consumables needed for 500 bombardments and is recommended for help in determining optimal conditions for a given cell type.
6. A space of approx 61 × 46 cm near a standard electrical outlet and vacuum source (or vacuum pump) is required. Such a space near a tissue culture hood and cell culture incubator is most convenient.
7. Standard low-cost microfuge (picofuge).

2.2. Solutions to Coat the Gold Microspheres With DNA

1. Solution of 50% water and 50% glycerol (sterile).
2. Aqueous solution of 1 M $CaCl_2$ (sterile).
3. Aqueous solution of 10 mM spermidine (sterile, free base, tissue culture grade).
4. Ethanol, 100 and 70% (high-performance liquid chromatographic or spectrophotometric grade).
5. Isopropanol (70%).

3. Methods

The basic principle of action is to propel microcarriers (gold or tungsten particles coated with cDNA) at the cells of interest in a manner that does not retard movement of the microcarriers. Thus, media must be aspirated from cells because liquid impedes microcarrier movement. Moreover, prior to bombardment, a partial vacuum is created inside the bombardment chamber to attenuate the slowing of microcarriers by air. Although one may expect aspiration of media and formation of a partial vacuum to adversely affect cultured cells, this is not the case if cells are exposed to these conditions for only the limited time required to bombard them with microcarriers. Thus, it is important to perform the actual bombardment (**Subheading 3.2., step 4**) briskly. With practice, this does not present a challenge. Note that the operator determines bombardment pressure not by using a regulator, but rather by selecting from an assortment of rupture discs that burst at a variety of pressures.

As might be intuitively expected, the disks rated at a low pressure are the flimsiest, and the ones rated at the higher pressures are increasingly stiff and thick. For bombardment, pressure builds from compressed helium until the rupture disk ruptures, propelling a second plastic disk called the macrocarrier, which has previously been coated with cDNA-coated microcarriers, toward a small metal screen (stopping screen) placed between the macrocarrier and the cultured cells to be transfected. The sudden stop of the macrocarrier disk allows microcarriers to be launched toward the target cells at high velocity. An overview of the bombardment procedure is outlined in **Fig. 1**. Variables that must be optimized include burst pressure (rupture disk choice) and the position (distance from stopping screen) of the target shelf, which will contain the cells to be bombarded. Once optimized for a given cell type, these variables change little for subsequent bombardments.

3.1. Prebombardment

1. Prepare stock of gold particles (microcarriers). Weigh 30 mg of 1 μm dry gold particles in a 1.5-mL microfuge tube. Add 1 mL 70% ethanol and vortex at maximal speed for 10–15 min. Allow particles to settle for 5 min and than briskly pellet them with a spin for 2–3 s in a microfuge. Discard the supernatant. Repeat three times except reduce vortexing and settling times to 1 min for subsequent ethanol washes. Following the final ethanol wash, add 500 μL sterile 50% glycerol and resuspend by vortexing. Microcarrier stock suspension can be stored at 4°C for up to 3 mo (*see also* **Notes 1** and **2**).

2. Cells can be cultured and plated as usual for electrophysiological recordings. We prepare SCG *(8)* and TG *(9)* neurons and subsequently plate them on coverglass (no. 1) chips using standard methods. Chips containing cultured cells are cultured in a standard 35-mm tissue culture dish containing medium. Bombardment of these cells is completed in these dishes. Because we find that neurons survive better when cultured at a high density, we plate dissociated cells onto a fairly small number of chips. This also allows moving the chips together in the center of the dish to catch as many of the bombarding gold particles as possible. At 1–3 d after bombardment, cells cultured on chips are transferred to the recording chamber for electrophysiological assay.

3. Check the helium supply in the tank and clean/sterilize equipment used in the bombardment chamber (rupture disk retaining cap, microcarrier launch assembly) and the consumables used in the bombardment (macrocarriers/macrocarrier holders) (*see* **Note 1**). We typically clean/autoclave the equipment and reusable material immediately following use and store until next use in the tissue culture room.

4. At some point well before the time of bombardment, prepare the consumables according to the PDS-1000 instructions, including preassembling and presterilizing the macrocarriers, transferring rupture disks to dishes, sterilizing a batch of stopping screens, and coating the washed microcarriers with DNA at a time more than 1 h before the planned bombardment. Coating is a critical part of the procedure as it is very important to avoid aggregation of the microcarriers since this may dramatically reduce efficiency of transfection (*see* **Notes 2** and **3**). Prior to aliquoting microcarriers to be coated, vortex microcarrier stock suspension for 5–15 min. Then, quickly aliquot 5 μL of the microcarriers into a fresh 1.5-mL microcentrifuge tube. While vortexing, quickly add 2–5 μg plasmid cDNA, 25 μL 1 M CaCl$_2$, and 40 μL 10 mM spermidine. Continue vortexing for 2 min, then allow

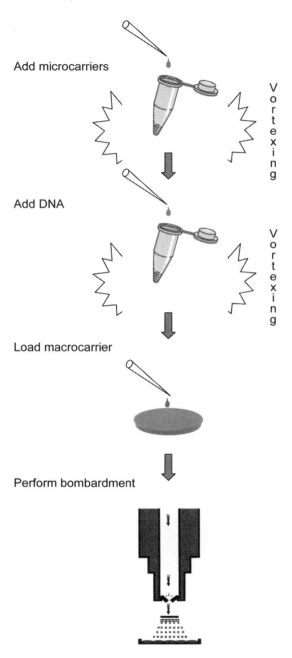

Fig. 1. Outline of the basic procedure of the transfection of cultured neurons using the biolistic particle delivery system (PDS)-1000/He.

microcarriers to settle for 1 min. Pellet coated particles by briskly centrifuging and remove and discard the supernatant. Add 150 μL 70% ethanol to the pellet without resuspending it and discard wash (ethanol supernatant) without disturbing the pellet. Repeat wash step with 150 μL 100% ethanol and gently resuspend coated microcarriers in 10–15 μL 100% ethanol. Do not vortex, but instead vigorously triturate coated microcarriers using a pipetteman. Note that ordinary "maxiprep" cDNA can be used to coat microcarriers, and multiple plasmids can be used together. We usually include a plasmid coding enhanced GFP (or a variant) as an optical marker for successful transfection. Although the final resuspension of cDNA-coated microcarriers can be in as little as 8 μL ethanol per bombardment, we prefer to resuspend in 15 μL ethanol per bombardment.

5. After coating microcarriers with cDNA, the 15 μL suspension is applied to the macrocarrier with one macrocarrier used per bombardment. Fill the bottom of a 100-mm culture dish with desiccant and place in a tissue culture hood; insert one macrocarrier into the macrocarrier holder and seat with the supplied plastic seating tool (this silly looking tool is actually very effective). Remember that sterile conditions must be maintained. For each bombardment, 15 μL microcarrier suspension is removed with a pipetteman and applied to the surface of the macrocarrier on the side away from its holder. It is best to apply the microcarrier suspension *immediately* after trituration. Quickly spread microcarrier suspension onto the macrocarrier. Because the pressure wave can only project the DNA that is coated on the area over the aperture of the macrocarrier holder, restrict coating to this area. Allow coated macrocarriers to dry in the desiccant-filled dish for 30–60 min.

3.2. Bombardment

1. The procedure is most convenient if the PDS-1000 is within reach of the tissue culture hood. Have fresh tissue culture medium ready in the hood and ensure that the vacuum is on. Insert a rupture disk into the presterilized retaining cap after dipping the rupture disk in 70% isopropanol to sterilize it. The choice of rupture disk determines the pressure of bombardment, which defines the spread pattern and depth of projection. We find empirically that the optimum pressure for transfection of SCG and TG neurons is 650 psi; however, we recommend that you determine the optimum pressure for your particular cell type. Ensure that only one rupture disk is placed in the retaining cap. If there is more than one disk, then the burst pressure will continue rising to exceed that expected to rupture the disk, which may adversely affect transfection. Hand-tighten the retaining cap containing the rupture disk to the end of the gas acceleration aperture and then further tighten using the torque wrench supplied with the equipment.

2. Place a stopping screen (presterilized) and the macrocarrier holder containing dried cDNA-coated microcarriers (facing down toward the stopping screen) into the fixed nest in the launch assembly (already inserted into the launch shelf), and hand-tighten the cover lid. Place the assembly in the top slot inside the bombardment chamber. It is important not to forget the stopping screen because, in its

absence, the macrocarrier will impact the cultured cells, leading to their wide-spread destruction.

3. The 35-mm culture dishes containing chips with cells to be transfected are placed on the target shelf, which can be inserted in any of the slots in the bombardment chamber. We use the uppermost slot. This, combined with bombardment pressure defined by rupture disk, sets the bombardment spread pattern. Remember that one goal is for cells to capture as many of the bombarding particles as possible. Place the vacuum to vent. You are now ready to perform the bombardment using the steps described next. These are to be performed briskly and in order.

4. Partly pull out the target shelf for access. Remove the dish of cells to be bombarded from the incubator and aspirate the medium from the dish. It is not necessary to wash cells at this point. You have about 1–2 min before the cells completely dry out, which is ample time for the bombardment. Immediately place the dish in the center of the target shelf, fully push the target shelf in, close and seal the bombardment door, press the vacuum button to the VAC position, and observe formation of the vacuum in the bombardment chamber on the chamber pressure gage.

   Most house vacuum systems will easily go to 15 mmHg. The point of the vacuum is to retard slowing of the bombarding particles by air. When the vacuum reaches the maximum characteristic of your system (this should only take a few seconds), briskly switch the vacuum switch to the HOLD position. Do not let the switch pause in the VENT position, or you will quickly vent away the vacuum that you have just created. The system will fire only when greater than 5 mmHg of vacuum is established, and the FIRE button should illuminate when at least 5 mmHg vacuum is reached.

   With the vacuum level stabilized in the bombardment chamber, press and hold the FIRE button to allow helium pressure to build inside the gas acceleration tube that is sealed by the selected rupture disk. You will hear a loud pop when the disk ruptures, which should be near the pressure (observed on the pressure gage on the top of the machine) rated for that disk (*see* **Note 4**). Immediately release the FIRE button and turn the vacuum switch to VENT.

   When the chamber has vented sufficiently to allow opening of the bombardment chamber door (this should take only a few seconds), remove the dish of cells, and immediately add fresh medium that is ready inside the hood. Place the cells in the incubator. Loosen the retaining cap with the torque wrench and dump out the burst rupture disk. Remove the launch assembly and unscrew its cover. Remove the macrocarrier holder and set aside for the next session. Remove the stopping screen and spent macrocarrier from the nest and discard. You are now ready to repeat the cycle for the next bombardment.

3.3. Postbombardment

1. After re-adding medium to cultured cells, the microcarriers can be easily observed with an ordinary tissue culture microscope. There should be a fairly dense lawn of particles visible. A cell with a particle visible in its nucleus will likely be transfected.

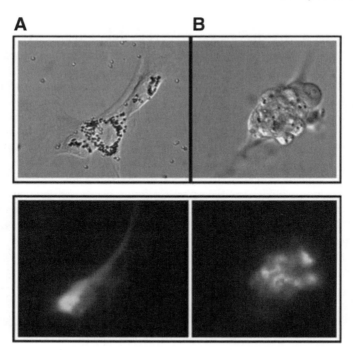

Fig. 2. Transfection of superior cervical ganglion cultures with enhanced green fluorescent protein. Top and bottom panels show transmitted light and fluorescent images, respectively, of (**A**) glial cells and (**B**) neurons. Details are given in the text.

2. After the bombardment session is completed, close the helium cylinder valve and chamber door. With more than 5 mmHg of vacuum in the chamber, remove the residual line pressure by pressing and holding the FIRE button. Vent any residual vacuum from the chamber by turning the vacuum button to VENT. Turn off the main vacuum supply and turn off the power to the machine.

3. The amount of time required for expression of the exogenous protein is variable for different proteins, but for ion channels and signaling proteins it is typically 1–2 d. The inclusion of cDNA coding for GFP is highly desirable as a means to identify successfully transfected cells. We often use a bicistronic vector that directs the simultaneous expression of enhanced GFP and a subcloned protein to ensure that green-fluorescing cells are also expressing the protein of interest. The chips that the cells were cultured and bombarded on can be directly used for electrophysiological analysis. The gold particles do not interfere with the patch clamp technique.

3.4. Transfection of Rat SCG and TG Neurons Using Biolistic PDS

1. Shown in **Fig. 2** is an SCG culture transfected with enhanced GFP. Transmitted light images are shown in the top panel; the bottom panel depicts fluorescent

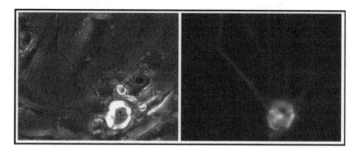

Fig. 3. A superior cervical ganglion neuron transfected with the phospholipase Cδ-pleckstrin homology-enhanced green fluorescent protein construct. Left image is the transmitted light micrograph, and on the right is a fluorescent image of the same field.

images of the same cells. **Figure 2A** shows a group of glial cells, and **Fig. 2B** shows a cluster of SCG neurons. Cells were isolated from SCG of 3- to 14-d-old male rats (Sprague Dawley) using the method of Bernheim et al. *(8)* and cultured for 2–4 d on 4 × 4 mm glass cover slip chips (coated with poly-L-lysine) at 37°C (5% CO_2). Fresh culture medium containing nerve growth factor (50 ng/mL) was added to the cells 3 h after plating. Cultured cells were transfected according to the protocol described in **Subheading 3.2.** Fluorescent microscopy was performed with an inverted Nikon (Tokyo, Japan) Eclipse TE300 microscope with an oil immersion, ×40, 1.30-numerical aperture objective. A Polychrome IV monochromator (TILL Photonics, Martinsreid, Germany) was used as the excitation light source, and an FITC HQ 96170M filter cube (Chroma Technology, Brattleboro, VT) was used for enhanced GFP imaging. Cells were excited at 470 nm, the fluorescence emission was collected by an IMAGO 12-bit cooled CCD camera, and images were stored with *TILLvisION* 4.01 software.

2. A similar technique was used to express in SCG and TG neurons an exogenous, optical reporter of phospholipase C (PLC) activity to study activation of PLC by G protein-coupled receptors. The reporter is a construct containing the pleckstrin homology (PH) domain of PLCδ fused with enhanced GFP. The PH domain of PLCδ binds to the substrate of PLC, the plasma membrane phospholipid phosphatidylinositol 4,5-bisphosphate (PIP_2) and the product of its hydrolysis, inositol trisphosphate (IP_3) *(10–13)*. Because the affinity of the PLCδ-PH construct to IP_3 is about 10-fold higher than to PIP_2 *(11–13)*, it serves as a useful reporter for PIP_2 hydrolysis by PLC.

In unstimulated cells, the cytosolic IP_3 concentration is low, and almost all the PLCδ-PH resides in the plasma membrane, bound to PIP_2. Optically, such localization results in the sharp and bright fluorescence of the cell contour (because it is a vertical projection of the considerable membrane area) with much weaker fluorescence in the middle of the cell (**Figs. 3** and **4**). However, after PLC activation, the PLCδ-PH increasingly translocates to the IP_3 that is accumulating in the

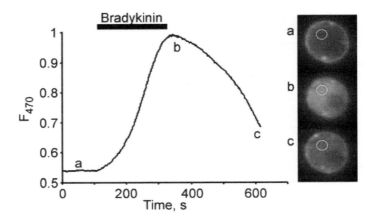

Fig. 4. Translocation of membrane-localized phospholipase Cδ-pleckstrin homology-enhanced green fluorescent protein (PLCδ-PH-EGFP) construct to the cytosol in a transfected trigeminal ganglion neuron reports activation of PLC by bradykinin stimulation. Plotted is cytosolic fluorescence of EGFP excited at 470 nM during bradykinin (200 nM) application (indicated by the bars). Insets on the right show images of the cell studied taken at the times indicated by letters. White circles indicate the "regions of interest" in the cytoplasm used for the fluorescence. Images (100 ms) were acquired every second.

cytosol, resulting in dimming of the fluorescence at the cell edges and an increase in the cytosolic fluorescence (**Fig. 4**).

This translocation of enhanced GFP fluorescence can be monitored as an optical readout of PLC activity. **Figure 3** shows an SCG neuron successfully transfected with the PLCδ-PH construct, and **Fig. 4** shows a translocation experiment made with a TG neuron. In these experiments, translocation of PLCδ-PH was induced by the stimulation of endogenous bradykinin B$_2$ receptors with 200 nM bradykinin. B$_2$ receptors activate PLC via the G$_{q/11}$ type of G$_α$ subunits. Translocation of PLCδ-PH was monitored as an increase in fluorescence excited at 470 nM (F$_{470}$) in the cytosolic region of the cell indicated by the white circles on the fluorescent images shown in the inset of **Fig. 4**. Images were taken at the time points indicated (**Fig. 4a,b,c**). Bath application of bradykinin induced robust translocation of the probe, providing a powerful tool to study G protein-coupled signaling in native mammalian neurons.

4. Notes

1. These methods and the manual of the PDS-1000 machine use a number of jargon words to describe the various components of the system. Many of these words are unintuitive and thus are defined here: *microcarriers* are the gold or tungsten microparticles coated with cDNA and used to bombard cells; *macrocarriers* are

the thin, amber-color circular disks that are coated with microcarrier suspension prior to bombardment (macrocarriers are projected downward until they impact the stopping screens); *macrocarrier holders* are round metal objects that look like large washers (macrocarriers are seated in these and placed, microcarriers facing down, in the nest of the launch assembly, over the stopping screens); and *rupture disks* are the small, circular disks that burst at a characteristic pressure (these are placed in the rupture disk retaining cap, which is hand-screwed and then torque wrench tightened over the gas acceleration tube at the top of the bombardment chamber).

2. A common pitfall of this method is overspinning, and resultant aggregation, of the microcarriers during washing and pelleting. The successful use of the system relies on the dispersion of the microcarriers as individual particles in suspension. Spinning them in a picofuge for 2–3 s or so is sufficient for pelleting, with longer spins easily causing aggregation in microcentrifuge tubes. Once aggregated in a tube, microcarriers are difficult to disperse.

3. Related to **Note 2** is the need for dispersion of the microcarriers into individual particles. Thus, the long (5-min) vortex times stated in this chapter are indeed necessary, especially those preceding aliquoting of the microcarriers for cDNA coating. Left sitting in a tube between bombardment sessions, the microcarriers will settle to the bottom, and a long vortex is required to redisperse them adequately. The use of a platform attachment is highly recommended. We solve problems of vortex wobbling during extended vortex times by placing it on a piece of Styrofoam.

4. If the disk does not rupture as the pressure rises past the burst pressure for that disk, then the usual reason is the accidental placement of more than one rupture disk in the retaining cap. This is very easy to do. Occasionally, a rupture disc will be defective and will not burst at the correct pressure. If this happens, then do not keep pressurizing the gas acceleration tube. Simply release the FIRE button, vent the bombardment chamber, place fresh medium in the dish containing cells to keep them alive, unscrew the retaining cap, and inspect the disk inside. Replace the disk as necessary and restart the process.

Acknowledgments

The work in this laboratory is funded by National Institutes of Health grant RO1 NS43394 (M. S. S.) and an American Heart Association postdoctoral fellowship 032512Y (N. G.).

References

1. Hille, B., Beech, D. J., Bernheim, L., Mathie, A., Shapiro, M. S., and Wollmuth, L. P. (1995) Multiple G-protein-coupled pathways inhibit N-type Ca^{2+} channels of neurons. *Life Sci.* **56,** 989–992.

2. Gamper, N. and Shapiro, M. S. (2003) Calmodulin mediates Ca^{2+}-dependent modulation of M-type K^+ channels. *J. Gen. Physiol.* **122,** 17–31.

3. Gamper, N., Stockand, J. D., and Shapiro, M. S. (2003) Subunit-specific modulation of KCNQ potassium channels by Src tyrosine kinase. *J. Neurosci.* **23,** 84–95.

4. Gamper, N., Reznikov, V., Yamada, Y., Yang, J., and Shapiro, M. S. (2004) Phosphatidylinositol 4,5-bisphosphate signals underlie receptor-specific $G_{q/11}$-mediated modulation of N-type Ca^{2+} channels. *J. Neurosci.* **24,** 10,980–10,992.

5. Patwardhan, A., Berg, K. A., Akopain, A. N., et al. (2005) Bradykinin-induced functional competence and trafficking of the delta opioid receptor in trigeminal nociceptors. *J. Neurosci.* **25,** 8825–8832.

6. Malin, S. A. and Nerbonne, J. M. (2000) Elimination of the fast transient in superior cervical ganglion neurons with expression of KV4.2W362F: molecular dissection of IA. *J. Neurosci.* **20,** 5191–5199.

7. Malin, S. A. and Nerbonne, J. M. (2001) Molecular heterogeneity of the voltage-gated fast transient outward K^+ current, I(Af), in mammalian neurons. *J. Neurosci.* **21,** 8004–8014.

8. Bernheim, L., Beech, D. J., and Hille, B. (1991) A diffusible second messenger mediates one of the pathways coupling receptors to calcium channels in rat sympathetic neurons. *Neuron* **6,** 859–867.

9. Price, T. J., Patwardhan, A., Akopian, A. N., Hargreaves, K. M., and Flores, C. M. (2004) Modulation of trigeminal sensory neuron activity by the dual cannabinoid-vanilloid agonists anandamide, *N*-arachidonoyl-dopamine and arachidonyl-2-chloroethylamide. *Br. J. Pharmacol.* **141,** 1118–1130.

10. Shaw, G. (1996) The pleckstrin homology domain: an intriguing multifunctional protein module. *Bioessays* **18,** 35–46.

11. Varnai, P. and Balla, T. (1998) Visualization of phosphoinositides that bind pleckstrin homology domains: calcium- and agonist-induced dynamic changes and relationship to myo-[3H]inositol-labeled phosphoinositide pools. *J. Cell. Biol.* **143,** 501–510.

12. Hirose, K., Kadowaki, S., Tanabe, M., Takeshima, H., and Iino, M. (1999) Spatiotemporal dynamics of inositol 1,4,5-trisphosphate that underlies complex Ca^{2+} mobilization patterns. *Science* **284,** 1527–1530.

13. Nash, M. S., Young, K. W., Willars, G. B., Challiss, R. A., and Nahorski, S. R. (2001) Single-cell imaging of graded $Ins(1,4,5)P_3$ production following G-protein-coupled-receptor activation. *Biochem. J.* **356,** 137–142.

II

METHODS FOR STUDYING
CHANNEL STRUCTURE–FUNCTION

4

Tertiary and Quaternary Structure Formation of Voltage-Gated Potassium Channels

John M. Robinson, Andrey Kosolapov, and Carol Deutsch

Summary

   Voltage-gated potassium channels are ubiquitous and critical for life. They must fold and assemble correctly and target to appropriate sites in the plasma membrane. Failure to do so can lead to inappropriate targeting or function and to pathology. The methods described here were developed to assess in which compartment tertiary and quaternary structure acquisition occurs. The experimental strategies involve identifying quaternary and tertiary interfaces, engineering a pair of cysteines into a cysteine-free voltage-gated potassium channel protein, using bifunctional crosslinking agents, and using an assay of the crosslinked products to determine folding/assembly events. A biogenic intermediate (i.e., nascent chain attached to transfer RNA and the ribosome) is used to probe events inside and at the exit port of the ribosomal tunnel.

   Key Words: Bifunctional cysteine reagents; biogenesis; crosslinking; pegylation; protein folding; ribosome; T1 domain; tetramerization; voltage-gated potassium channels.

1. Introduction

   Tetrameric voltage-gated potassium (Kv) channels originate as monomeric peptides attached to ribosomes (nascent peptides). Synthesis and targeting of the nascent peptide-ribosome complex to the endoplasmic reticulum (ER) membrane is followed by assembly and integration of the peptide into the bilayer. Included in these steps is formation of the tertiary and quaternary structures of the channel. Previous studies have indirectly implicated the ER as the site of Kv channel folding (*1–5*). However, only recently has direct evidence for quaternary (*6,7*) and tertiary (*8,9*) structure formation been obtained for the recognition domain (T1) of Kv channels. The T1 domain is a cytoplasmic N-terminal sequence that is highly conserved among Kv channels and is responsible for

From: *Methods in Molecular Biology, vol. 337: Ion Channels: Methods and Protocols*
Edited by: J. D. Stockand and M. S. Shapiro © Humana Press Inc., Totowa, NJ

subfamily-specific coassembly of subunits *(5,10,11)*. This chapter focuses on two biogenic events: tertiary folding of the T1 domain in the monomer and tetramerization of the T1 domains.

Regarding tertiary folding of the T1 monomer, we have developed a biochemical microassay to probe formation of the folded intramolecular interface of monomeric T1 in Kv1.3. This assay is independent of a measurement of tetramer formation and uses (1) bismaleimides to crosslink pairs of cysteines engineered into an internally folded interface of the monomeric T1 domain and (2) a gel shift strategy that extends current pegylation techniques *(12)*. This approach allows us to detect intramolecularly crosslinked Kv1.3 monomers *(8,9)*. Specifically, we have introduced pairs of cysteines into the T1 Kv1.3 monomer based on the crystal structure of the virtually identical T1 in Kv1.1a *(13)*.

Regarding quaternary structure formation of the T1 domain, we use similar strategies, except the cysteines are engineered at the T1-T1 intersubunit interface, and the resultant crosslinked products are multimeric species identified as dimer, trimer, and tetramer on a gel. In both tertiary and quaternary structure assays, we have generated biogenic intermediates that remain attached to the ribosome. However, mature Kv channel protein can likewise be analyzed.

These methods, previously described for Kv channel folding/assembly, can similarly be applied to any soluble protein or cytosolic domains for which a high-resolution structure is available.

2. Materials

2.1. Construct Design, Transcription, and Translation

1. *RasMol* computer program (http://www.umass.edu/microbio/rasmol/).
2. QuikChange Site-Directed Mutagenesis Kit (Stratagene, La Jolla, CA) and the relevant sense and antisense oligonucleotides (Invitrogen, Carlsbad, CA).
3. pSp64/Kv1.3 vector; XL-1-Blue Supercompetent (Stratagene).
4. QIAfilter Plasmid Midi Kit (Qiagen, Valencia, CA); 1X Tris/ethylenediaminetetraacetic acid buffer solution pH 7.4 or 8.0 (Fisher), depending on the restriction enzyme requirements. Appropriate restriction enzymes (NEB, Beverly, MA) (*see* **Note 1**).
5. Riboprobe transcription system with SP6 RNA polymerase (Promega, Madison, WI) (*see* **Note 2**).
6. Translation reagents: canine pancreatic microsomal membranes (ER-derived; Promega); nuclease-treated rabbit reticulocyte lysate (Promega); 1 mM amino acid mixture minus methionine (Promega); 40 U/µL ribonuclease (RNase) inhibitor (Promega); DEPC-treated sterile water (Fisher); and [^{35}S]methionine, 10 µCi/µL Express (Dupont/NEN Research Products, Boston, MA) (*see* **Notes 2** and **3**).

2.2. Tertiary and Quaternary Crosslinking

1. Buffer A: Dulbecco's phosphate-buffered saline without CaCl$_2$ and MgCl$_2$ (Invitrogen) at pH 7.3 and supplemented with 4 mM MgCl$_2$. Store at 4°C.

2. Buffer B: 50 mM NaCl, 1 mM MgCl$_2$, and 20 mM HEPES and pH 7.3. Store at 4°C.
3. Crosslinking agent: 250 mM stock solution ortho-phenyldimaleimide (o-PDM) (Sigma) dissolved in 99.9% dimethyl sulfoxide (DMSO) (Sigma) and stored at room temperature (*see* **Note 4**).
4. NuPAGE Sample Reducing agent (Invitrogen; or use dithiothreitol [DTT]) (*see* **Note 5**).
5. 2-Mercaptoethanol (Fisher).
6. Microcentrifuge II (Fisher).
7. Sucrose cushion: 0.5 M sucrose, 100 mM KCl, 50 mM HEPES, and 5 mM MgCl$_2$ at pH 7.5. Store at 4°C. NuPAGE Sample Reducing Agent is added fresh before use.
8. 1.5 mL Beckman Polyallomer microfuge tubes (Beckman Coulter, Fullerton, CA), Beckman TLA 100.3 rotor, and a Beckman Optima TLX ultracentrifuge.
9. Sodium dodecyl sulfate (SDS) 10% stock solution (Invitrogen).
10. Methoxy-polyethylene glycol maleimide (PEG-MAL) and methoxy-polyethylene glycol thiol (PEG-SH) (Nektar Therapeutics, Huntsville, AL) (*see* **Note 6**): 40 mM stock solutions diluted in buffer A or B (*see* **step 1** or **2**). Make fresh each experiment.
11. 2500:1 solution of acetone:12 N HCl.

2.3. Gel Electrophoresis and Fluorography (see Note 7)

1. Sample preparation: NuPAGE Sample Reducing Agent, loading buffer such as NuPAGE LDS sample buffer (4X) (Invitrogen), and double-distilled water (ddH$_2$O) added to the isolated protein sample.
2. Electrophoresis performed using NuPAGE system and precast 10, 12, or 4–12% Bis-Tris gels 1.0 mm thick, 10-well gels (Invitrogen).
3. Running buffer: NuPAGE MOPS SDS running buffer (20X) (Invitrogen).
4. Antioxidant: NuPAGE antioxidant (Invitrogen).
5. Fixation solution: 1500 mL methanol, 1200 mL ddH$_2$O, 300 mL acetic acid.
6. To enhance ^{35}S fluorography, gels are soaked in Amplify (Amersham, Arlington Heights, IL).
7. Gels are dried on a Bio-Rad model 583 Gel Dryer (Bio-Rad, Hercules, CA) using a Savant Refrigerated Condensation Trap and a Savant Gel Pump GP110 (Global Medical Instrumentation, Ramsey, MN).
8. Gel quantitation: Molecular Dynamics PhosphorImager (Amersham Biosciences, Piscataway, NJ) and Molecular Dynamics Storage Phosphor Screen (Amersham Biosciences) (*see* **Note 8**).

3. Methods

3.1. General Procedures

Both assays for tertiary and quaternary formation begin by using the *RasMol* program (or an equivalent program that can display pdb files) to scrutinize

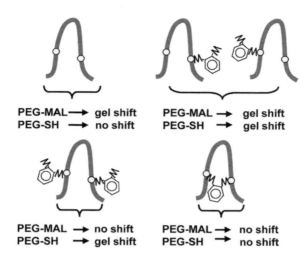

PEG-MAL⟶ gel shift   PEG-MAL ⟶ gel shift
PEG-SH ⟶ no shift   PEG-SH ⟶ gel shift

PEG-MAL ⟶ no shift   PEG-MAL ⟶ no shift
PEG-SH ⟶ gel shift   PEG-SH ⟶ no shift

Fig. 1. Strategy for the tertiary folding assay. Monomeric protein is shown as a thick curved line. Ortho-phenyldimaleimide is indicated by its chemical structure. Engineered cysteines at a putative intramolecular interface are shown as circles. The predicted results on treating each monomer with the indicated pegylating reagent are shown below each monomeric species. (Reproduced from **ref. 8** with permission from ASBMB, Inc.)

crystal structures that are homologous to our protein and find pairs of interface residues that come within 3–10 Å of each other. We replace these residues with cysteines, which contain a reactive thiol group at the γ position. In the folded structure, these cysteines can then be crosslinked with a bifunctional crosslinking reagent (e.g., PDM) that contains two maleimide groups, each of which is capable of covalently binding one thiol group.

In the case of tertiary folding, three criteria must be met to choose an appropriate pair of engineered cysteines *(8)*. First, the residues must be far enough apart in the primary sequence to avoid nonspecific crosslinking. Second, the pair must be within 5–10 Å in the folded T1 monomer and therefore within crosslinking distance. Third, the pair must be on the surface of the folded protein and therefore presumably nondisruptive and accessible to crosslinking reagents. Intramolecularly crosslinked (folded) protein can be distinguished from noncrosslinked (unfolded) protein by a mass-tagging strategy. Mass tagging is accomplished by modification of the PDM-treated protein with PEG-MAL or PEG-SH as shown in **Fig. 1**. Addition of PEG-MAL or PEG-SH (pegylation) shifts the protein molecular mass by 10 kDa or more. A free peptidyl-thiol group can be labeled with PEG-MAL *(12)*, and a free peptidyl-maleimide can be labeled with PEG-SH *(8,9)*.

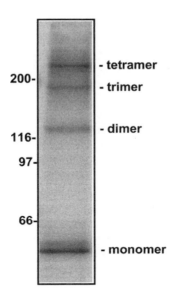

Fig. 2. Crosslinking of intersubunit cysteines in the quaternary folding assay. Protein was translated in a rabbit reticulocyte lysate in the presence of ^{35}S-methionine and microsomal membranes and treated with ortho-phenyldimaleimide (500 μM) to form monomers, dimers, trimers, and tetramers. Numbers to the left of the gel represent molecular weight standards in kilodalton. Cysteines were introduced into a cysteine-free background to ensure specific crosslinking. (Reproduced from **ref. 7** with permission from Elsevier, Inc.)

In the case of quaternary folding, the second and third criteria for the tertiary folding assay must also be met. An additional criterion applies to oligomeric structures containing more than two subunits: the pair should be on *different* intersubunit interfaces, thus permitting formation of dimers, trimers, and tetramers and not just dimers *(6)*. Crosslinked multimers are easily distinguished from each other on a gel (**Fig. 2**; *see also* **refs. 6** and 7). We use standard methods of mutagenesis, bacterial transformation, plasmid DNA preparation, restriction enzyme analysis, cRNA transcription, and in vitro protein translation. The following steps are common to both tertiary and quaternary folding experiments:

1. All cysteine mutations are introduced using a QuikChange Site-Directed Muta-genesis Kit and the relevant sense and antisense oligonucleotides. pSP64/Kv1.3 plasmid is amplified by transforming XL-1-Blue Supercompetent cells. cDNA purification is carried out using a QIAfilter Plasmid Midi Kit. cDNA is dissolved at approx 1 $\mu g/\mu L$ in Tris/ethylenediaminetetraacetic acid buffer solution. All mutant cDNAs are sequenced in the region of the mutation.

2. 20 µg cDNA are linearized by the appropriate restriction enzyme, followed by standard phenol/chloroform purification, and used to make cRNA.

3. Capped cRNA is transcribed from the linearized cDNA template using the Riboprobe In Vitro Transcription System with SP6 RNA polymerase according to the Promega technical manual.

4. Proteins are translated in vitro using rabbit reticulocyte lysate and microsomal membranes (when needed) according to the Promega Protocol and Application Guide. Preparation for the translation begins by removing all components from the −80°C freezer and placing them in an ice bucket. Allow all reagents to thaw completely before use. Next, for a 25-µL translation, add the reagents to a 1.5-mL Eppendorf tube (*see* **Note 3**) at room temperature in the following order: 1.2 µL DEPC H_2O, 0.5 µL amino acid mixture minus methionine, 1.0 µL RNase inhibitor, 1.8 µL microsomal membranes (*see* **Note 9**), 2.0 µL ^{35}S methionine, 17.5 µL rabbit reticulocyte lysate, and 1.0 µL cRNA, 0.2 µg/µL. It is important that the reaction does not begin until all reagents have been added. To ensure this, add each reagent to the wall of the Eppendorf tube so that none are in contact. After the lysate has been added, spin down the mixture briefly (2–3 s) in a microcentrifuge to collect everything at the bottom, then return tubes to the bench and add the cRNA to the mixture. Incubate for 2 h at 30°C for full-length constructs (released from ribosome) and 1 h at 22°C for biogenic intermediates (still attached to ribosome).

3.2. Tertiary Folding

3.2.1. Experimental Assay

1. Set up 50-µL translation reaction in a 1.5-mL Eppendorf tube (*see* **Notes 2** and **3**) for each peptide to be examined in the folding assay. You need to process three separate samples (tubes 1, 2, 3) for each translation. Two translations may be conveniently assayed simultaneously, which requires a six-tube rotor (**steps 4** and **7**).

2. While the translation reaction finishes, prepare the following solutions: for dilution buffer, aliquot 0.5 mL buffer A to 1.5-mL tubes. Add 2 µL reducing reagent to each tube, vortex, and store tubes on ice (*see* **Note 10**). In a separate 1.5-mL tube, aliquot enough sucrose cushion for the whole experiment. Add NuPAGE Sample Reducing Reagent at a 1:500 ratio, vortex, and aliquot 120 µL into high-speed 1.5-mL centrifuge tubes (*see* **Subheading 2.2.**, **step 8**). Store all tubes on ice. You need three tubes with buffer A and three tubes with sucrose cushion for each translation.

3. After the translation reaction is finished, add 10–15 µL of translation reaction to each tube containing dilution buffer. Gently vortex each sample. Carefully layer the sample onto the sucrose cushion in the correspondingly numbered tube. Mark sample tubes derived from the same translation as 1, 2, and 3. Store samples on ice.

4. Centrifuge (Beckman Optima TLX Ultracentrifuge, Beckman TLA 100.3 rotor) all samples from **step 3** through the sucrose cushion at 208,000*g* for 20 min at

4°C (*see* **Note 9**) to isolate ribosome-attached nascent peptides and remove unwanted SH groups. While waiting for next step, prepare 250 mM PDM stock in DMSO solution. Vortex it vigorously and store at room temperature (*see* **Note 4**).

5. Promptly remove supernatant with suction (yellow pipet tip attached to vacuum) from each tube. Be careful not to disturb the pellet, which is not visible. Add 0.5 mL ice cold buffer A to each tube and vortex gently (*see* **Note 11**). Add 1 μL PDM stock solution to tubes 2 and 3 and vortex them gently until visible particles of PDM/DMSO disappear. No additions are made to sample 1 as sample 1 is the control for determining cysteine reactivity. Incubate samples 1–3 for 30 min on ice. While the samples are incubating, prepare 0.5 M 2-mercaptoethanol stock solution: add 36 μL 2-mercaptoethanol to 964 μL DEPC-treated water; mix and store the stock solution on ice.

6. Samples 1 and 2 will be examined for reactive cysteines. Sample 3 will be examined for reactive maleimide groups in PDM-modified protein. Unreacted PDM in sample 2 is quenched by adding 10 μL 2-mercaptoethanol stock and incubating it for 5 min at room temperature and 5 min on ice. Treat sample 1 (control) the same way as sample 2. Leave sample 3 on ice without treatment; otherwise, quenching will modify reactive peptidyl-maleimide groups.

7. Centrifuge all samples at 208,000g for 20 min at 4°C (*see* **Note 9**). While centrifuging, prepare protein denaturing solutions for the next step. Prepare 1 mL 1% SDS solution (denaturing solution) using 10% SDS and buffer A in two separate 1.5-mL tubes (one to be used for samples 1 and 2, the other for sample 3). Add 20 μL 2-mercaptoethanol stock to 1 mL 1% SDS solution prepared for samples 1 and 2 only. Sample 3 will be denatured in the absence of reducing agents for the reason described in **step 6**.

8. Protein denaturing increases the pegylation (**step 9**) efficiency. Remove the supernatant as described in **step 5**. Add 50 μL denaturing solution with 10 mM 2-mercaptoethanol to samples 1 and 2. Add 50 μL denaturing solution without reducing agent to sample 3. Vortex and briefly (1–3 s) spin tube contents down with microcentrifuge to collect any sample that may have been retained on the tube walls. Incubate samples for 30 min at room temperature. While the samples are incubating, prepare 40 mM PEG-MAL and PEG-SH stock solutions using buffer A. Vortex the samples vigorously and store the PEG solutions on ice.

9. Assay samples 1 and 2 for reactive cysteines with PEG-MAL and sample 3 for reactive peptidyl-maleimide with PEG-SH. Add 50 μL PEG-MAL or PEG-SH stock solution to the appropriate tube. The final concentration of PEG-MAL or PEG-SH is 20 mM. Vortex, spin tube briefly (1–3 s), and incubate samples at 4°C for 2 h.

10. Acetone precipitate the protein by adding 900 μL acetone-HCl solution to each tube (90% of final volume). Vortex and store samples at −20°C overnight (*see* **Note 12**).

11. Recover protein by centrifuging (Eppendorf centrifuge 5415C at 16,000g, i.e., maximum speed) the sample for 30 min at 4°C. Remove the supernatant and allow the tubes to air-dry at room temperature for 1 h.

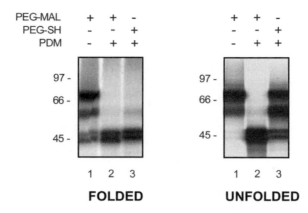

Fig. 3. Crosslinking of intramolecular cysteines in the tertiary folding assay. Nascent peptide (attached to ribosomes) was translated in a rabbit reticulocyte lysate in the presence of ^{35}S-methionine and microsomal membranes and treated as follows: **Left gel**: the sample shown in lane 1 was treated with methoxy-polyethylene glycol maleimide (PEG-MAL); the sample in lane 2 was treated first with ortho-phenyl-dimaleimide (*o*-PDM) and then PEG-MAL; the sample in lane 3 was treated first with *o*-PDM and then PEG-SH. **Right gel**: All samples were treated first with SDS to denature the nascent peptide and then treated as described for the **left gel**. Numbers to the left of the gel represent molecular weight standards in kilodalton. Numbers to the right of the gel indicate singly (1), doubly (2), and unpegylated (0) peptide. All samples were treated with RNase (20 µg/mL) prior to loading on the gel to remove the peptidyl-transfer RNA bands. The upper band of the doublet at 43 kDa is caused by core glycosylation in the ER membranes of this protein; the lower one is unglycosylated protein. Core glycosylation is not readily discerned at higher molecular weights. Cysteines were introduced into a cysteine-free background to ensure specific crosslinking. (Reproduced from **ref. 8** with permission from ASBMB, Inc.)

12. Prepare samples for gel electrophoresis by adding 25–30 µL premixed loading buffer (NuPAGE LDS 4X sample buffer) to each tube (*see* **Subheading 3.3.1.**, **step 5** for content and recipe; *see* **Notes 12** and **13**). After the gel is processed (*see* **Subheading 2.3.**, **step 7**), radioactive protein is detected as a ladder of distinctive bands (pegylation ladders), quantified using a phosphorimaging system (*see* **Subheading 2.3.**, **step 8**), and analyzed for P_{fold} (*see* **Subheading 3.2.2.**).

3.2.2. Analysis of Pegylation Ladders

For any given construct, radioactive protein incubated with PEG-MAL or PEG-SH is detected as distinct bands on NuPAGE gels for folded (**Fig. 3, left gel**) and unfolded (**Fig. 3, right gel**) protein. The bands are quantified using phosphorimaging, and the data are analyzed using the following equations. For more details, refer to **refs. 8** and **9**.

For each lane j of the gel, the fraction of total protein molecules with i pegylated cysteines is calculated as $Wj(i) = \mathrm{cpm}(i)/\Sigma\mathrm{cpm}(i)$, where $\mathrm{cpm}(i)$ is the counts per minute in the ith bin. For example, in **Fig. 3**, in which each Kv construct has two cysteines, i ranges from 0 to 2. If each cysteine is assumed to label to the same final extent, then the fraction Fj of individual cysteines pegylated in the jth lane is $\Sigma iWj(i)/N$, where N is the total number of cysteines in the protein molecule. For the gels in **Fig. 3**, F_1 is the fraction of individual cysteines labeled by PEG-MAL (**lane 1**). F_2 is the fraction of individual cysteines labeled by PEG-MAL after treatment with PDM (**lane 2**). F_3 is the fraction of individual cysteines that has reacted with both PDM and PEG-SH (**lane 3**).

By comparing the labeling in denatured (SDS pretreatment) vs nondenatured protein, we can estimate the crosslinking efficiency as follows. After SDS pretreatment (**Fig. 3**, **right gel**), $F_{\mathrm{PDM\text{-}SDS}}$ is the fraction of individual cysteines labeled with PDM, given by $F_{\mathrm{PDM\text{-}SDS}} = (F_1 - F_2)/F_1$. From the same gel, $F_3 = F_{\mathrm{PDM\text{-}SDS}}P_{\mathrm{PEG\text{-}SH}}$, where $P_{\mathrm{PEG\text{-}SH}}$ is the probability that an individual cysteine labeled with PDM has reacted with PEG-SH. Thus, $P_{\mathrm{PEG\text{-}SH}} = F_3/\{(F_1 - F_2)/F_1\}$.

Using this estimate of $P_{\mathrm{PEG\text{-}SH}}$, we now can determine the probability of a pair of cysteines being crosslinked by PDM in the absence of SDS pretreatment (**Fig. 3**, **left gel**). As above, the fraction of individual cysteines labeled by PDM is $F_{\mathrm{PDM}} = (F_1 - F_2)/F_1$. The fraction of available free maleimides after PDM labeling in this case is $F_{\mathrm{fMAL}} = F_3/P_{\mathrm{PEG\text{-}SH}}$, where F_3 is determined from **lane 3** in the right gel of **Fig. 3**, and $P_{\mathrm{PEG\text{-}SH}}$ is estimated as described above from SDS-pretreated channels. Finally, the probability of a pair of cysteines being crosslinked by PDM, P_{xlink}, is a measure of the extent of folding. Thus, $P_{\mathrm{fold}} = P_{\mathrm{xlink}} = F_{\mathrm{PDM}} - F_{\mathrm{fMAL}}$. One-way analysis of variance (ANOVA) or a Student t test is used to determine whether differences in P_{fold} values are statistically significant.

In the case of folding assays carried out on different segments of the Kv channel protein, which may be in different compartments (e.g., inside the ribosome vs in the cytosol), a modification of this analysis and experimental data is required. Another set of equations (*see* **ref. 9**) is used. These equations require the additional determination of Fj for single cysteine constructs for each pair of cysteines.

3.3. Quaternary Folding

3.3.1. Experimental Assay

1. During the time of the translation reaction, prepare a 250-mM stock solution of PDM in DMSO and leave it at room temperature until the translation is finished.
2. Once the reaction has finished, in a second 1.5-mL Eppendorf tube (*see* **Note 3**) dilute 5 µL translation mixture into 500 µL buffer B (*see* **Note 14**) and add 1 µL PDM stock solution (0.5 mM final) to crosslink. Incubate on ice for 5 min (*see* **Note 15**).

3. Quench the mixture with 5 µL NuPAGE Sample Reducing Agent (500 mM stock) to a final concentration of 5 mM. Note that the reducing agent concentration is 10X greater than the PDM concentration to ensure complete quenching. Incubate the sample at room temperature for 15 min.
4. During the quenching step, prepare a stock solution of sucrose cushion by adding NuPAGE Sample Reducing Agent to a final concentration of 1 mM (1 µL stock per 500 µL solution). Add 120 µL of this final sucrose cushion to each of the 1.5-mL Beckman centrifuge tubes. After quenching, pipet the crosslinked mixture on top of the cushion. The mixture should be added slowly so the two layers do not mix. Centrifuge in a Beckman TLX ultracentrifuge for 7 min at 106,000g and 4°C (*see* **Note 9**).
5. Remove supernatant by suction (*see* **Note 16**). Add 2.4 µL NuPAGE Sample Reducing Agent, 6 µL NuPAGE sample buffer, and 15 µL ddH$_2$O and denature at 70°C for 15 min.

3.3.2. Analysis of Quaternary Folding

The crosslinking experiments using Kv1.3 protein typically yield four distinct bands on a gel, corresponding to monomer, dimer, trimer, and tetramer of the Kv channel protein (*see* **Fig. 2**; *see also* **refs. 6** and **7**). To determine the fraction of the total protein that each band represents, we divide the counts per minute (cpm) of each band by the total counts per minute for all four bands.

4. Notes

1. Our restriction enzymes function best in pH 7.4 buffer; however, other enzymes may require a different pH to linearize plasmid DNA. Refer to restriction enzyme literature to determine optimal conditions.
2. All transcription and translation materials are stored at –80°C.
3. We recommend using 1.5-mL RNase-free tubes (EMSCO, Philadelphia, PA) for translation and for all solutions made for the assay, but not for high-speed centrifugation.
4. PDM must be made fresh for each experiment because the maleimide groups will hydrolyze in solution over time. Although the crosslinking reaction is done on ice, the stock solution must be left at room temperature because the DMSO will freeze on ice.
5. NuPAGE Sample Reducing Agent, DTT, or 2-mercaptoethanol are added fresh to solutions for each experiment.
6. PEG-MAL is sensitive to moisture because it can be hydrolyzed; PEG-SH is sensitive to oxidation. Both reagents should be stored under inert atmosphere according to company specifications. We buy them in either 1- or 5-g bottles and then aliquot those bottles into 1.5-mL Eppendorf tubes and store them desiccated and under inert conditions at –20°C.
7. Other comparable reagents and polyacrylamide gel electrophoresis systems can be used instead of NuPAGE. We use NuPAGE because it gives the best resolution.

8. We quantitate our gels directly using a Molecular Dynamics PhosphorImager, which detects counts per minute that are not necessarily visualized in autoradiograms exposed for 16–30 h. Thus, some bands, at the level of 5–10% of the protein, are not visible but are detected by PhosphorImaging.

9. Translation reactions for some tertiary folding experiments may be made in the absence of microsomal membranes if monomeric cytoplasmic domains are to be studied in the absence of membrane-anchored domains. If the protein to be assayed is membrane-free, then centrifuge at 208,000g for 20 min. If the protein to be assayed is membrane-incorporated (translation with membranes), then centrifuge at 106,000g for 7 min.

10. Keep samples and stock solutions on ice unless otherwise specified. Such tactics will minimize thiol oxidation and maleimide hydrolysis.

11. The folding assay, as applied to predenatured protein (negative control), has to be modified as follows because the denatured protein cannot be recovered by centrifugation (**Subheading 3.2.1., step 7**): in **Subheading 3.2.1., step 5**, add 25 μL 1% SDS in buffer A to the protein pellets; incubate the sample at room temperature for 30 min to denature the protein. Add 25 μL buffer A to each tube to dilute SDS, thereby preventing SDS precipitation during the subsequent PDM-labeling step, which is performed on ice. Add 1 μL 25 mM PDM stock solution to samples 2 and 3. In **Subheading 3.2.1., step 6**, use 1 μL 0.5 M 2-mercaptoethanol stock solution to treat samples 1 and 2. Omit **Subheading 3.2.1., steps 7** and **8**. Before proceeding to **Subheading 3.2.1., step 9**, increase the SDS concentration in each sample to 1% by adding 2.7 μL 10% SDS solution. Use 55 μL 40 mM PEG-MAL and PEG-SH stock solutions, respectively, to pegylate samples 2 and 3. The changes cited are made to keep the folding assay conditions for native and predenatured peptides as identical as possible.

12. Samples can be stored for a few days at –20°C after **Subheading 3.2.1., steps 10** and **12** and at room temperature after **step 11**.

13. If the peptidyl-transfer RNA band is not desired in the final gel, then treat the sample with RNase (20 μg/mL) at the end of **Subheading 3.2.1., step 12**. Add 1–2 μL RNase solution (stock is 500 μg/mL) to the loading buffer, DTT, and water (final volume approx 24 μL) and incubate for 15 min at room temperature.

14. We typically dilute 5 μL translation per 500 μL phosphate-buffered saline buffer; however, for a stronger signal, add more translation mixture (i.e., 10, 15, 20 μL) and keep all other quantities the same.

15. These conditions are sufficient to achieve maximal final labeling. We have carried out this reaction for up to 1 h and obtain the same amount of crosslinking that we do after 5 min.

16. After removing the supernatant, we typically do not see a pellet because there is so little protein. Do not think you have lost your protein because you cannot see it.

Acknowledgments

This work was supported by National Institutes of Health grant GM 52302 (C. D.), American Heart Association grant-in-aid (C. D.), and National

Research Service Award HL-07027 (A. K.). John M. Robinson developed the quarternary assay. Andrey Kosolapov developed the tertiary assay.

References

1. Rosenberg, R. L. and J. E. East. (1992) Cell-free expression of functional *Shaker* potassium channels. *Nature* **360,** 166–169.
2. Deal, K. K., Lovinger, D. M., and Tamkun, M. M. (1994) The brain Kv1.1 potassium channel: in vitro and in vivo studies on subunit assembly and posttranslational processing. *J. Neurosci.* **14,** 1666–1676.
3. Nagaya, N. and Papazian, D. M. (1997) Potassium channel α and β subunits assemble in the endoplasmic reticulum. *J. Biol. Chem.* **272,** 3022–3027.
4. Schulteis, C. T., Nagaya, N., and Papazian, D. M. (1998) Subunit folding and assembly steps are interspersed during *Shaker* potassium channel biogenesis. *J. Biol. Chem.* **273,** 26,210–26,217.
5. Shen, N. V., Chen, X., Boyer, M. M., and Pfaffinger, P. (1993) Deletion analysis of K$^+$ channel assembly. *Neuron* **11,** 67–76.
6. Lu, J., Robinson, J. M., Edwards, D., and Deutsch, C. (2001) T1–T1 interactions occur in ER membranes while nascent Kv peptides are still attached to ribosomes. *Biochemistry* **40,** 10,934–10,946.
7. Robinson, J. M. and Deutsch, C. (2005) Coupled tertiary folding and oligomerization of the T1 domain of Kv channels. *Neuron,* **45,** 223–232.
8. Kosolapov, A. and Deutsch, C. (2003) Folding of the voltage-gated K$^+$ channel T1 recognition domain. *J. Biol. Chem.* **278,** 4305–4313.
9. Kosolapov, A., Tu, L., Wang, J., and Deutsch, C. (2004) Structure acquisition of the T1 domain of Kv1.3 during biogenesis. *Neuron* **44,** 295–307.
10. Li, M., Jan, Y. N., and Jan, L. Y. (1992) Specification of subunit assembly by the hydrophilic amino-terminal domain of the *Shaker* potassium channel. *Science* **257,** 1225–1230.
11. Xu, J., Yu, W., Jan, Y.-N., Jan, L.-Y., and Li, M. (1995) Assembly of voltage-gated potassium channels. Conserved hydrophilic motifs determine subfamily-specific interactions between the α-subunits. *J. Biol. Chem.* **270,** 24,761–24,768.
12. Lu, J. and Deutsch, C. (2001) Pegylation: a method for assessing topological accessibilities in Kv1.3. *Biochemistry* **40,** 13,288–13,301.
13. Kreusch, A., Pfaffinger, P. J., Stevens, C. F., and Choe, S. (1998) Crystal structure of the tetramerization domain of the *Shaker* potassium channel. *Nature* **392,** 945–948.

5

Biophysical Approach to Determine the Subunit Stoichiometry of the Epithelial Sodium Channel Using the *Xenopus laevis* Oocyte Expression System

Farhad Kosari, Shaohu Sheng, and Thomas R. Kleyman

Summary

The amiloride-sensitive epithelial Na^+ channel (ENaC) is typically composed of three structurally related subunits termed α, β, and γ. We describe methods to determine the functional subunit stoichiometry of ENaC based on a biophysical approach that was first introduced in 1991 to determine the subunit stoichiometry of a voltage-gated K^+ channel. The strategy is to analyze channel sensitivity to a specific blocker when various mixtures of block-sensitive and blocker-insensitive subunits are coexpressed in a heterologous expression system. Details related to the expression of wild type and mutant ENaCs in *Xenopus* oocytes and the examination of blocker sensitivity by two-electrode voltage clamp, as well as analysis of data are provided.

Key Words: Amiloride; ENaC; two-electrode voltage clamp.

1. Introduction

Several laboratories have investigated the subunit stoichiometry of the epithelial Na^+ channel (ENaC) *(1–5)*. Here, we describe a biophysical approach that was first adopted by MacKinnon to estimate the subunit stoichiometry of a voltage-activated potassium channel *(6)*. An ENaC is composed of three structurally related subunits, termed α, β, and γ. The number of individual subunits within a channel complex can be deduced by the approach described in this chapter. The strategy to determine subunit stoichiometry is based on the generation of mutant ENaC subunits that result in a large change in the sensitivity of channels to a specific blocker compared to wild-type (WT) channels. Coexpression of both WT and mutant

From: *Methods in Molecular Biology, vol. 337: Ion Channels: Methods and Protocols*
Edited by: J. D. Stockand and M. S. Shapiro © Humana Press Inc., Totowa, NJ

subunits will result in heteroligomeric channels with varying sensitivities to channel blockers. Analyses of the effects of inhibitors on channel activity allow determination of subunit stoichiometry (6–8).

As an example, consider a channel that is a homo-oligomer, and its subunit stoichiometry is to be determined. Assume that the channel is blocked by compound X, and that a specific mutation within the channel renders the channel insensitive to this compound. To determine subunit stoichiometry, equal amounts of WT and mutant subunits are expressed in a heterologous expression system, such as *Xenopus* oocytes. If there is only one subunit in a channel complex, then half of the channels expressed will be drug sensitive (i.e., WT), and half of the channels will be drug insensitive. Addition of a saturating concentration of the inhibitor would block half of the current carried by the channel. If there are two subunits/channel, then there are four possible combinations: WT/WT, WT/mutant, mutant/WT, and mutant/mutant. In this case, 25% of the current would be blocked by the inhibitor. If the number of subunits/channel complex is greater than two, then the percentage of current blocked by the inhibitor in cells in which equal amounts of WT and mutant subunits are expressed will decrease in a predictable manner (*see* **Table 1**).

This analysis of subunit stoichiometry rests on several assumptions: (1) assembly of channels containing WT or mutant subunits is a random event; (2) a single drug-sensitive subunit confers blocker insensitivity to the heteroligomeric channel; and (3) other than differential sensitivities to inhibitors, functional properties of mutant and WT channels are similar (6,7). Amiloride is a prototypic inhibitor of ENaC (9). Mutations at a putative amiloride-binding site, αS583C, βG525C, and γG538C, were used to determine ENaC subunit stoichiometry as channels with one of these mutant subunits differ from WT ENaC in sensitivity to channel blockers. For example, both $\alpha\beta$G525Cγ and $\alpha\beta\gamma$G538C channels were relatively insensitive to amiloride (K_i increased by approx 1000-fold). Although αS583C$\beta\gamma$ channels were sensitive to amiloride (K_i for amiloride increased only sixfold), these channels were blocked by sulfhydryl reactive reagents, including (2-aminoethyl) methanethiosulfonate (MTSEA) and (2-[trimethylammonium] ethyl) methanethiosulfonate (MTSET).

To estimate the stoichiometry of an ENaC subunit, WT and the corresponding mutant cRNAs of that subunit are mixed at fixed ratios and coexpressed with the other two WT subunits in oocytes. The stoichiometry of the subunit can then be estimated by analyses of titration curves for reversible channel blockers, such as amiloride, or by analyses of remaining currents following an irreversible channel blocker, such as MTSEA or MTSET.

2. Materials

Reagents were purchased from vendors listed in this subheading or from Sigma Chemical Company (St. Louis, MO).

Table 1
Prediction of the Response in Channels Expressed in Oocytes Injected With a 1:1 Mixture of Blocker-Sensitive (○) and Blocker-Insensitive (●) Subunit cRNAs

Stoichiometry	Channel species	Species population (%)	Predicted remaining current for the individual species (%)	Predicted remaining whole cell current (%)
1	○	50	100	50
	●	50	0	
2	○○	25	100	25
	●●	25	0	
	○●	25	0	
	●○	25	0	
3	○○○	12.5	100	
	●●●	12.5	0	12.5
	○●●	12.5	0	
	●○●	12.5	0	
	●●○	12.5	0	
	●○○	12.5	0	
	○●○	12.5	0	
	○○●	12.5	0	

2.1. Solutions

1. All solutions are sterilized by filtering with 0.22 micron cellulose acetate membrane filters (Corning Inc., Corning, NY) and stored at 4°C or room temperature.
2. Following injections, oocytes are incubated in modified Barth's saline (MBS; *see* ref. *10*): 88 mM NaCl, 1 mM KCl, 2.4 mM NaHCO$_3$, 15 mM HEPES, 0.3 mM Ca(NO$_3$)$_2$, 0.41 mM CaCl$_2$, 0.82 mM MgSO$_4$, pH 7.2. The buffer is supplemented with 10 μg/mL penicillin, 10 μg/mL streptomycin sulfate and 100 μg/mL gentamycin sulfate. MBS was stored at 4°C and used within a week.
3. Bath solution for two-electrode voltage clamp (TEV) measurements (*see* **Note 1**): 100 mM sodium gluconate, 2 mM KCl, 1.8 mM CaCl$_2$, 10 mM HEPES, 5 mM BaCl$_2$, and 10 mM tetraethylammonium chloride at pH 7.2. The solution is stored at 4°C and used within 1 mo.
4. Pipet solution: glass pipets are filled with 3 M KCl.
5. MTSEA is purchased from Toronto Research Chemicals Inc. (Toronto, Ontario, Canada). MTSEA powder is added into the bath solution to make 0.5 mM working solution immediately prior to use.

6. Amiloride solutions: amiloride-containing solutions can be conveniently prepared by diluting 1 m*M* amiloride stock solution with the bath solution. The stock is made by dissolving amiloride in the bath solution. Optionally, a higher concentration of amiloride stock can be prepared in dimethyl sulfoxide.

7. Anesthetic solution: tricane (ethyl 3-aminobenzoate methanesulfonate salt) is freshly prepared at 0.18% in water.

8. Calcium-free solution: 88 m*M* NaCl, 2 m*M* KCl, and 5 m*M* HEPES at pH 7.2.

9. Collagenase in calcium-free solution: 1.5–2.0 mg/mL collagenase type IV or I is prepared in calcium-free solution.

2.2. Instruments

1. Two-electrode voltage clamp (TEV): a TEV 200 amplifier (Dagan Corp., Minneapolis, MN) and DigiData 1200 analog-to-digital (A/D) converter (Axon Instrument, Union City, CA) were connected to a PC via an industry standard architecture (ISA) expansion card for data acquisition (*see* **Note 2**). Newer DigiData series from Axon has an small computer systems interface (SCSI) interface. Two Ag/AgCl electrode pellets from World Precision Instruments (WPI, Sarasota, FL) or Warner Instruments Inc. (Hamden, CT) attached to a silver wire are connected to the bath reference head stage and placed in the oocyte chamber.

2. Oocyte chamber: oocyte chambers for TEV experiments are purchased from Warner Instruments.

3. Perfusion: we use the gravity perfusion method for delivery of reagents (MTSEA or amiloride; *see* **Note 3**).

4. Stereomicroscopes: for both injection and TEV, we use SZ-60 microscopes with fiber-optic light sources from Olympus (*see* **Note 2**).

5. Micromanipulators: MM-33 micromanipulators (Fine Science Tools Inc., Foster City, CA) are used to move the pipet electrodes (*see* **Note 2**).

6. Glass capillaries and pipet puller: injection pipets are pulled from 3.5-in. glass capillaries (1.14 mm od, 0.53 mm id) from Drummond Scientific Company (Broomall, PA) with a Flaming/Brown Micropipette Puller (model P-87) from Sutter Instrument Company (Novato, CA). The fine top is broken mechanically to give a tip diameter of 20–30 µm. Recording pipets are pulled from borosilicate glass capillaries (1.5 mm od, 0.84 mm id; WPI) with a multiple-step program.

3. Methods

3.1. Expression of WT and Mutant Channels in Xenopus *Oocytes*

3.1.1. Oocyte Isolation and Preparation

1. Adult female *Xenopus laevis* frogs are purchased from Xenopus One Incorporated (Dexter, MI); Xenopus Express Incorporated (Plant City, FL); or Nasco (Fort Atkinson, WI) and maintained in water tanks at 18–22°C on a 12-h light/dark cycle. Frogs are anesthetized by immersion in 0.18% tricane in cold water for 20 min. Ovaries are removed and placed in Petri dishes filled with MBS. Harvested ovary lobes can be treated immediately for injection or maintained at 18°C for several days prior to collagenase treatment.

2. Separate oocytes into small clumps (10–20 oocytes each) with fine forceps and transfer to new MBS.
3. Wash the oocytes twice with calcium-free solution.
4. Transfer oocytes to a 15-mL Corning tube filled with 10 mL calcium-free solution containing 1.5–2.0 mg/mL collagenase type IV (Sigma).
5. Slowly swirl the oocytes on a horizontal shaker for approx 1 h. After 1 h, check oocytes under the microscope every 5 min to determine the extent of defolliculation. Stop the process when most oocytes are free of connective tissues, which are readily identified by visible blood vessels.
6. Remove collagenase solution and wash five times with calcium-free solution. Transfer oocytes to a new Petri dish with MBS and place the oocytes dish at 18°C.
7. Let oocytes recover for a minimum of several hours prior to injection of cRNAs. We find that oocytes survive injection better if allowed to recover overnight.

3.1.2. Preparation of cRNA

1. Synthesize mutant and WT cRNAs from cDNA templates using a commercial RNA synthesis kit. We generally use the mMessage mMachine In Vitro Transcription Kit (Ambion Inc., Austin, TX). Synthesized cRNA should be purified by phenol/chloroform extraction followed by ethanol precipitation or by passing through RNA purification columns guaranteed nuclease free. Store 2- to 5-μL aliquots at −80°C.
2. Determine cRNA concentrations with an ultraviolet spectrophotometer or denaturing agarose gel (*see* **Note 4**).
3. cRNA mixing: WT and mutant cRNAs for α, β, and γ ENaC subunits are mixed with specific ratios according to experimental design and coinjected for optimal expression of the channels. Calculate the dilutions for the mutant and the WT subunits for injection of 1–3 ng of each subunit in 50 nL volume. To estimate stoichiometry of an ENaC subunit, mix a fixed ratio (1:1, 1:4, and so on) of the WT and the corresponding mutant cRNA. Add equal quantities of the other two WT subunit cRNAs. All dilution and mixing must be done on ice to avoid degradation. Store mixed cRNAs on ice for immediate injection or at −80°C for future injections.

3.1.3. Oocyte Injection

1. Select approx 50 healthy oocytes for each group and place them on a mesh mounted on the bottom of a 5-cm Petri dish filled with MBS.
2. Thaw and place mixed cRNAs on ice.
3. Fill a pulled injection pipet with mineral oil and attach it to a NanoJect (Drummond Scientific Co., Broomall, PA) microinjector. Move the injector plug to position for taking RNA samples. Transfer 1–2 μL mixed cRNAs on a piece of clean parafilm and draw RNA solution into the pipet by retreating the plug. Stop drawing when nearly all RNA solution is taken in to avoid air bubble formation inside the pipet.
4. Inject equal amounts (1–3 ng) of each of the three subunit cRNAs in a total volume of 50 nL into each oocyte (*see* **Note 4**).

5. Repeat **steps 3** and **4** for each additional group of oocytes.
6. Incubate injected oocytes in separate 35-mm dishes or in a six-well Cell Culture Cluster (Corning Inc., Corning, NY) containing MBS solution at 18°C. Change MBS daily. Dead cells should be removed promptly to minimize damage to healthy oocytes.

3.2. Two-Electrode Voltage Clamp

3.2.1. Determination of α-Subunit Stoichiometry

1. TEV is generally performed 1–3 d following cRNA injection *(11)*. Create a clamping protocol with the help of the software manual.
2. Pull recording pipets and fill them with 3 *M* KCl. Attach pipets to pipet holders and test pipet resistances. Pipets with a resistance of 0.5–5 MΩ are suitable for TEV.
3. Place an oocyte in the oocyte chamber.
4. Insert the electrodes into the oocyte. Start perfusion of the bath solution. Flow rates will vary according to the oocyte perfusion chamber used.
5. Clamp the oocyte to −100 mV (with reference to the bath, which is 0) for 480. The current measured at 400 ms is used for data analysis.
6. Prepare fresh bath solution with 0.5 m*M* MTSEA (*see* **Notes 5** and **6**).
7. Perfuse the bath with 0.5 m*M* MTSEA bath solution for 2–3 min and record the currents, as in **step 5**.
8. Perfuse the bath for 5–6 min with regular bath solution, followed by bath solution supplemented with 100 µ*M* amiloride (*see* **Note 7**). Measure the amiloride-insensitive current.
9. Determine the amiloride-sensitive current by subtracting the amiloride-insensitive current (**step 8**) from the starting current (**step 5**).
10. Determine the percentage of the remaining current after MTSEA perfusion by subtracting the amiloride-insensitive current (**step 8**) from the remaining current (**step 7**) and divide by the amiloride-sensitive current (**step 9**; *see* **Note 6**).

3.2.2. Determination of β- and γ-Subunit Stoichiometries

1. The large differences in the sensitivities of WT αβγ and αβG525Cγ ENaCs, and differences of αβγ and αβγG542C to amiloride are used to determine β- and γ-subunit stoichiometry. Here, the procedure for determining β-subunit stoichiometry is discussed. Stoichiometry for the γ-subunit can be determined using a similar approach.
2. Inject four groups of 20 oocytes with mouse ENaC (mENaC) cRNAs for (1) αβγ; (2) αβG525Cγ; (3) αβmixγ (βmix, 1:1 β to βG525C); and (4) αβmixγ (βmix, 1:4 β to βG525C). Incubate the injected oocytes in MBS at 18°C.
3. Perform TEV as in **Subheading 3.2.1.**, **steps 1–5** to measure the starting currents.
4. Measure the currents in the presence of increasing concentrations of amiloride.
5. Perfuse the bath for 5–6 min with the bath solution supplemented with high concentrations of amiloride (5 m*M*). Measure the amiloride-insensitive current.
6. Determine the amiloride-sensitive current by subtracting the amiloride-insensitive current (**step 5**) from current measurements in **steps 3** and **4**.

3.3. Data Analysis

3.3.1. α-Subunit Stoichiometry (see Notes 8 and 9)

1. Enter the data from experiments above in a spreadsheet.
2. Estimate the predicted responses to MTSEA for α-subunit stoichiometries (N_a) of 1, 2, or 3. In making these predictions, the basic model assumptions described in the introductory remarks are assumed to be valid. For models of subunit stoichiometries of 2 and 3, set predicted hybrid channel currents to the appropriate values with smallest possible χ^2 error when compared with experimental results (*see* **Note 10**).
3. Determine if the predicted remaining currents (PRCs) for α-subunit stoichiometries of 1, 2, or 3 are statistically different from the measured remaining currents by Student *t* test.
4. Our results were most consistent with an α-subunit stoichiometry of 2 *(3)*. In our experiments, the measured remaining currents of oocytes injected with a 1:1 ratio of WT to α-mutant channel cRNAs were statistically different from PRCs for $N_a = 1$ and $N_a = 3$ ($p < 0.01$). But, the PRC for an $N_a = 2$ was not significantly different from the measured remaining current ($p > 0.3$). For the oocytes injected with a 4:1 ratio of WT to α-mutant channel cRNAs, the PRC for an $N_a = 3$ was statistically different from the measured remaining currents ($p < 0.05$). However, the PRCs for $N_a = 1$ and $N_a = 2$ were not significantly different from measured remaining currents ($p = 0.11$ and $p > 0.3$, respectively).

3.3.2. β- and γ-Subunit Stoichiometries

1. Enter the data for the measured remaining amiloride-sensitive currents of oocytes from each group in spreadsheets. Plot the remaining currents that are normalized to the basal currents against amiloride concentrations.
2. Obtain inhibitory constant (K_i) and Hill coefficient for the αβγ and αβG525Cγ mENaCs by fitting the dose–response data with the Hill equation, $I_R = K_i^n/(C^n + K_i^n)$, where I_R is the remaining current normalized to the control current; K_i is the inhibitory constant; C is the blocker concentration; and n is the Hill coefficient.
3. There are several ways to estimate subunit stoichiometry from the data gathered from the previously mentioned experiments. Analyses with all three methods will increase the confidence of data interpretation.

3.3.2.1. METHOD A. COMPARISON OF THE OBSERVED RESPONSE WITH PREDICTED RESPONSES FOR DIFFERENT STOICHIOMETRY NUMBERS

1. Estimate the predicted responses to amiloride for channels containing one or two β-subunit (N_b). A similar estimation can be made for γ-subunit stoichiometry (N_g) of either 1 or 2. In making these predictions, the basic model assumptions noted in the introductory remarks are assumed to be valid. If channels are composed of two β subunits, then three distinct populations of channels will be present: (1) WT channels, (2) fully β-mutant channels, and (3) channels that have both a WT and a β-mutant subunit (i.e., hybrid channels). Similarly, if channels are com-

posed of two γ subunits, then there will be three populations of channels. In models for subunit stoichiometry of two β or γ subunits, the inhibition constant for amiloride ($K_{i\text{-hyb}}$), Hill coefficient (HC_{hyb}), and single-channel conductance (γ_{hyb}) for hybrid channels were obtained by minimizing the χ^2 error of the predicted response for a stoichiometry of 2 to the experimental data (*see* **Note 11**).

2. Estimate the stoichiometries of β and γ subunits by comparing the measured and the PRCs following amiloride perfusion for subunit stoichiometries of 1 and 2 (*see* **Figs. 5** and **6** in **ref. 3**). For statistical verifications, one can employ likelihood ratio analysis.

3. Our results were most consistent with β- and γ-subunit stoichiometries of 1. The amiloride titration data points fit the PRCs for β- and γ-subunit stoichiometries of 1 much better than β- and γ-subunit stoichiometries of 2. We also performed likelihood analyses for statistical validation. These analyses suggested that β- and γ-subunit stoichiometries of 1 are much more likely to produce the experimental results than β- and γ-subunit stoichiometries of 2 *(3)*.

3.3.2.2. METHOD B. PLOTTING THE DOSE–RESPONSE DATA WITH EQUATIONS THAT PREDICT SUBUNIT STOICHIOMETRY (ADOPTED FROM REF. 6)

The amiloride dose–response data are transformed according to the following equation:

$$(1/\ln(f)) \times \ln(U_{\text{mix}}/U_{\text{bmut}}) = N - (1/(f)) \times \ln(1 - R/U_{\text{mix}})$$

where U_{mix} represents unblocked response from oocytes injected with a (1:1 or 1:4) mixture of WT and mutant β-subunit mENaC cRNAs together with WT α and γ mENaC cRNAs. U_{bmut} denotes unblocked response of αβG525Cγ mENaCs. R is the unblocked response caused by all species containing one or more WT β-subunits; f represents the effective cRNA ratio, which includes a correction for the differences in single-channel conductances between WT and mutant mENaCs, assuming a stoichiometry of 1. As amiloride concentration increases, R diminishes, and the left side of the equation approaches N, the β-subunit stoichiometry. Accordingly, γ-subunit stoichiometry can be estimated similarly using γG542C. As shown in **Figs. 5** and **6** in **ref. 3**, using methods A and B, there likely are only one β and one γ subunit in αβγ mENaC complex.

3.3.2.3. METHOD C. FITTING THE AMILORIDE RESPONSES OF WT (AMILORIDE-SENSITIVE) AND MUTANT (AMILORIDE-INSENSITIVE) CHANNELS TO THE SUM OF TWO HILL EQUATIONS TO DEFINE β- AND γ-SUBUNIT STOICHIOMETRY

1. The mutant channel (αbG525Cγ) is about 1000-fold less sensitive to amiloride than WT (αβγ) *(12)*. For the channels with a mixture of WT and G525C β subunits, the dose-response for amiloride block is expected to be biphasic because of the coexistence of amiloride-sensitive and amiloride-insensitive β subunit. If β-subunit stoichiometry is one, then there will be either WT or mutant channels with no hybrid channels containing both WT and mutant β subunits. For this case,

the amiloride dose–response curve for a mixed population of channels with either WT or mutant β subunits is predicted to have two components: a high affinity one with a K_i similar to that for WT channels and a low affinity one with a K_i similar to that of the mutant channel.

2. Fit the dose–response data with the two-component equation:

$$I_R = 1 - \{AC/(C + K_1) + BC/(C + K_2)\}$$

where I_R and C have the same meaning as above; K_1 and K_2 are the Ki values for the two components in the dose-response curve; A and B are the fractions of each component, and their sum equals the total decrease of I_R in the dose-response curve. If the summated two Hill equations can satisfactorily fit the dose-response data for a mixed population of channels with two K_i values comparable to those of WT (αβγ) and mutant (βG525Cγ) variants and fractions (A and B) of the high-affinity component and the low-affinity component correspond to the fractions of the WT β subunit and mutant β subunits in the mixed channels, then the β-subunit stoichiometry is likely 1. Otherwise, there may be more than one β subunit in αβγ ENaCs (*see* **Note 11**). Using this method, Firsov and coworkers concluded that the stoichiometry for both β and γ subunits is 1 for rat αβγ ENaC (*1*).

4. Notes

1. A sodium gluconate solution is used for TEV to minimize potential current contamination from endogenous chloride channels in *Xenopus* oocytes. The potassium channel blockers $BaCl_2$ and TEA are included in the bath solution to minimize current contamination from endogenous K^+ channels in oocytes.

2. Comparable TEVC amplifiers, manipulators, and microscopes and their accessories are commercially available from several companies.

3. Any of several commercially available perfusion systems may be used to better control perfusion efficiency.

4. Accurate measurement of cRNA concentrations is critical. We recommend using a spectrophotometer that does not require dilution of cRNA samples, such as ND-1000 (NanoDrop, Wilmington, DE). Furthermore, we recommend that prior to taking spectrophotometric measurements, in vitro synthesized cRNA be purified using commercially available reagents, such as the RNeasy Kit (Qiagen, Valencia, CA) to remove free nucleotides.

5. MTS reagents have short half-lives and hydrolyze rapidly in aqueous solutions. It is important to prepare fresh MTSEA (or MTSET) solution just prior to its use.

6. MTSEA is membrane permeant at neutral pH (*13*). The membrane impermeant positively charged MTSET is also suitable for determining α-subunit stoichiometry and offers several advantages for these studies. For example, MTSET does not block WT channels.

7. A wash period was included in our protocols to deplete the bath solution of MTSEA to ensure that amiloride-induced block of channels occurred without possible interference by free MTSEA.

8. There are limitations to using a biophysical approach for determining subunit stoichiometry, as this method relies on several assumptions discussed in the intro-

duction that may not hold for a given channel. These key assumptions should be experimentally confirmed when possible.

9. The methods described here allow for determination of functional subunit stoichiometry of ENaC. The subunit stoichiometry determined with this method reflects active channels. Interestingly, it has been reported that only a small portion of channels in oocyte plasma membranes are active when α, β, and γ ENaCs are coexpressed *(14)*. It is not known whether inactive channels at the plasma membrane have a subunit stoichiometry identical to that of active channels.

10. MTSEA blocks WT channels. Our experience suggests that block of WT channels is modest (<10%) when a concentration of 0.5 mM MTSEA is used and oocytes are exposed to MTSEA for only a brief (3 min) period of time.

11. Although difficult, it is possible to judge with reasonable confidence the blocker sensitivity of hybrid channels with both WT and mutant subunits by examining the shifts of dose-response curves from the WT-mutant coinjected oocytes when compared to oocytes expressing either WT or mutant channels *(1,6)*.

References

1. Firsov, D., Gautschi, I., Merillat, A. M., Rossier, B. C., and Schild, L. (1998) The heterotetrameric architecture of the epithelial sodium channel (ENaC). *EMBO J.* **17**, 344–352.

2. Snyder, P. M., Cheng, C., Prince, L. S., Rogers, J. C., and Welsh, M. J. (1998) Electrophysiological and biochemical evidence that DEG/ENaC cation channels are composed of nine subunits. *J. Biol. Chem.* **273**, 681–684.

3. Kosari, F., Sheng, S., Li, J., Mak, D. O., Foskett, J. K., and Kleyman, T. R. (1998) Subunit stoichiometry of the epithelial sodium channel. *J. Biol. Chem.* **273**, 13,469–13,474.

4. Berdiev, B. K., Karlson, K. H., Jovov, B., et al. (1998) Subunit stoichiometry of a core conduction element in a cloned epithelial amiloride-sensitive Na⁺ channel. *Biophys. J.* **75**, 2292–2301.

5. Staruschenko, A., Medina, J. L., Patel, P., Shapiro, M. S., Booth, R. E., and Stockand, J. D. (2004) Fluorescence resonance energy transfer analysis of subunit stoichiometry of the epithelial Na⁺ channel. *J. Biol. Chem.* **279**, 27,729–27,734.

6. MacKinnon, R. (1991) Determination of the subunit stoichiometry of a voltage-activated potassium channel. *Nature* **350**, 232–235.

7. Ferrer-Montiel, A. V. and Montal, M. (1996) Pentameric subunit stoichiometry of a neuronal glutamate receptor. *Proc. Natl. Acad. Sci. USA* **93**, 2741–2744.

8. Liman, E. R., Tytgat, J., and Hess, P. (1992) Subunit stoichiometry of a mammalian K⁺ channel determined by construction of multimeric cDNAs. *Neuron* **9**, 861–871.

9. Kleyman, T. R. and Cragoe, E. J., Jr. (1988) Amiloride and its analogs as tools in the study of ion transport. *J. Membr. Biol.* **105**, 1–21.

10. Ceriotti, A. and Colman, A. (1995) mRNA translation in *Xenopus* oocytes. *Methods Mol. Biol.* **37**, 151–178.

11. Stuhmer, W. (1998) Electrophysiologic recordings from *Xenopus* oocytes. *Methods Enzymol.* **293**, 280–300.

12. Schild, L., Schneeberger, E., Gautschi, I., and Firsov, D. (1997) Identification of amino acid residues in the α, β, and γ subunits of the epithelial sodium channel (ENaC) involved in amiloride block and ion permeation. *J. Gen. Physiol.* **109,** 15–26.

13. Holmgren, M., Liu, Y., Xu, Y., and Yellen, G. (1996) On the use of thiol-modifying agents to determine channel topology. *Neuropharmacology* **35,** 797–804.

14. Firsov, D., Schild, L., Gautschi, I., Merillat, A. M., Schneeberger, E., and Rossier, B. C. (1996) Cell surface expression of the epithelial Na channel and a mutant causing Liddle syndrome: a quantitative approach. *Proc. Natl. Acad. Sci. USA* **93,** 15,370–15,375.

6

Spectroscopy-Based Quantitative Fluorescence Resonance Energy Transfer Analysis

Jie Zheng

Summary

The combination of green fluorescent protein mutants and fluorescence resonance energy transfer (FRET) forms a powerful tool for ion channel studies. A key to successful application of green fluorescent protein-based FRET is to reliably separate the FRET signal from various non-FRET fluorescence emissions that coexist in any experimental system. This chapter introduces a FRET quantification method that is based on fluorescence spectroscopic microscopy. Application of this "spectra FRET" method to both the confocal imaging of *Xenopus* oocytes and the epifluorescence imaging of culture cells is described. The fluorescence intensity ratio measurement, a complementary non-FRET method for identifying the channel subunit stoichiometry, is also discussed.

Key Words: Confocal microscopy; epifluorescence microscopy; fluorescence intensity ratio; fluorescence resonance energy transfer; protein–protein interaction; spectroscopy; stoichiometry.

1. Introduction

Fluorescence resonance energy transfer (FRET) has been widely used to detect both intra- and intermolecular interactions in living cells *(1–4)*. FRET reports the proximity of two fluorophores *(5–7)*. Light energy absorbed by a donor fluorophore is transferred to a nearby acceptor fluorophore with an absorption spectrum that overlaps the emission spectrum of the donor. The efficiency of energy transfer falls off with the sixth power of the distance between the donor and acceptor molecules, making FRET an extremely sensitive reporter of proximity. A combination of the discovery of green fluorescent protein (GFP)-based FRET pairs, for example, the cyan fluorescent protein (CFP) and yellow fluorescent protein (YFP) pair *(8,9)*, and the fast

From: *Methods in Molecular Biology, vol. 337: Ion Channels: Methods and Protocols*
Edited by: J. D. Stockand and M. S. Shapiro © Humana Press Inc., Totowa, NJ

advance in fluorescence microscopy technology has resulted in the enthusiastic adoption of FRET approaches to a wide spectrum of biological studies of molecule colocalizations, specific protein–protein interactions, and conformational rearrangements *(10)*.

Application of the GFP-based FRET to ion channel studies has yielded new information on specific interactions between ion channel proteins and modulatory proteins *(11–15)* and factors *(16)*, as well as between channel subunits *(17–22)*. These studies benefited greatly from the covalent linkage of the fluorescent protein to the channel protein, which ensures stoichiometric fluorescence labeling. The relatively fast maturation of the fluorescent protein also allows tracking of channel subunits during assembly and trafficking in live cells *(23,24)*.

A major drawback of the fluorescent protein, however, is the rather large size of this molecule, which limits the spatial resolution of GFP-based FRET approaches. The minimum distance between the donor and acceptor chloroforms, located at the center of the fluorescent protein, is expected to be close to R_0, the characteristic distance for 50% FRET efficiency *(9)*. Hence, it is unlikely to achieve high degrees of energy coupling. In addition, FRET occurs between the fluorophore pairs, which are somewhat away from the sites on the channel protein to which they are attached. As a result, caution should be taken in interpreting results from GFP-based FRET experiments in terms of the spatial resolution.

Reliable quantification of the FRET efficiency is a key to the success of any FRET experiment. Various quantitative FRET approaches have been developed *(25,26)*. In this chapter, a spectroscopy-based, quantitative FRET measurement method is discussed. This "spectra FRET" method provides several advantages *(18)*. First, it is insensitive to the relative fluorescence intensities between the donor fluorophore and the acceptor fluorophore, allowing comparison across cells with different expression levels. Second, it allows accurate separation of the donor emission from the acceptor emission range (the bleed-through problem) as well as correction for the direct excitation of the acceptor by the excitation light intended for exciting the donor (the cross-excitation problem). Finally, it provides an internal check for the linearity of the recording system.

Equipments discussed here are suitable for experiments using enhanced cyan fluorescent protein (eCFP) and enhanced yellow fluorescent protein (eYFP), the most commonly used GFP mutant FRET pair. The protocol can, however, be easily adapted to other fluorophores with minor modifications of the equipment configuration. In addition to the spectra FRET approach, a complementary non-FRET fluorescence approach for identification of the channel subunit stoichiometry is also discussed.

Fluorescence recordings have been frequently carried out with *Xenopus* oocytes or culture cells. The oocyte expression system has been widely used in ion channel studies because of their ability to robustly express most channel types and because of the low level of native membrane proteins, especially native channel proteins. A specific advantage of the oocyte expression system for FRET experiments is that the relative expression levels of the CFP-tagged subunit and the YFP-tagged subunit can be reliably controlled by the amount of cRNAs injected. Hence, one can easily optimize the ratio of CFP-tagged subunits to YFP-tagged subunits for specific FRET measurements (*see* **Note 1**).

Fluorescence imaging of oocytes, however, has been proven to be nontrivial, partially because of the strong autofluorescence from the intracellular milieu. The autofluorescence problem can be overcome by imaging oocytes with a confocal microscope *(18)*. This rather atypical recording configuration—using high-resolution confocal microscopy on large, featureless oocytes—yields a special benefit: Only fluorescence signals from the oocyte surface membrane are collected under such a recording condition *(13,15,17,18)*. This unique feature, especially when combined with the intrinsically low autofluorescence level of the dark side (the animal pole) of oocytes, makes the oocyte a nice system for fluorescence studies of ion channels.

Culture cells, on the other hand, are convenient to use because one transfection reaction yields a large number of fluorescent cells. Unlike oocytes, culture cells are transparent to fluorescence emission, allowing observation of intracellular distribution and trafficking of channel subunits that are tagged with fluorescence proteins. In addition, fluorescence recordings from culture cells can be conveniently done using a regular epifluorescence microscope. In this chapter, fluorescence recording protocols specifically designed for oocytes and culture cells are provided.

2. Materials

2.1. Equipment

1. Confocal microscope (Leica TCS NT/SP, Leica Corp., Wetzlar, Germany).
2. Epifluorescence microscope (Olympic IX81, Olympus USA, Melville, NY).
3. Spectrograph (Acton SpectraPro215, Roper Scientific Inc., Tucson, AZ).
4. Charge-coupled device (CCD) camera (Roper cascade 128B, Roper Scientific).

2.2. Software

1. *MetaMorph* (for control of the epifluorescence microscope and the CCD camera, imaging, and off-line data analysis) (Molecular Devices Corp., Sunnyvale, CA).
2. Microsoft *Excel* (used in conjunction with *MetaMorph* for tabulating spectra during data analysis) (Microsoft Corp., Redmond, WA).
3. *SpectroPro* (for control of the spectrograph) (Roper Scientific).
4. The *LCS* software (for control of the Leica confocal microscope) (Leica).

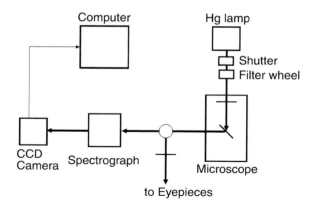

Fig. 1. Schematic diagram of the epifluorescence spectroscopy setup. Light path is shown in bold lines.

2.3. Epifluorescence Spectroscopic Imaging

1. Human embryonic kidney epithelial cell line (HEK) 293 cells expressing CFP-tagged channel subunits.
2. HEK 293 cells expressing YFP-tagged channel subunits.
3. HEK 293 cells coexpressing CFP-tagged channel subunits and YFP-tagged channel subunits.
4. Cell imaging solution: 130 mM NaCl, 5 mM MgCl$_2$, 2 mM CaCl$_2$, 1 mM EGTA, and 5 mM HEPES pH 7.4.
5. Recording chamber (Nunc Lab-Tek).
6. Immersion oil.

2.4. Confocal Fluorescence Imaging

1. *Xenopus laevis* oocytes expressing CFP-tagged channel subunits.
2. Oocytes expressing YFP-tagged channel subunits.
3. Oocytes coexpressing CFP- and YFP-tagged channel subunits.
4. Oocyte solution: 130 mM NaCl, 3 mM HEPES, 0.2 mM ethylenediamine-tetraacetic acid (EDTA) at pH 7.2 adjusted with 1 M N-methyl-glutamine.
5. Recording chamber (Nunc Lab-Tek).

3. Methods

3.1. FRET Imaging of Culture Cells Using Epifluorescence Microscopy

3.1.1. Spectroscopic Imaging of Culture Cells

The following protocol is based on the fluorescence spectroscopy setup as shown in **Fig. 1.** It is built around a fully computer-controlled Olympus IX81 epifluorescence microscope. For excitation, a mercury light source is used in conjunction with a shutter. A filter wheel that houses a set of neutral density filters is added to the excitation light path to control the light intensity. Two filter cubes, each containing an excitation filter and a dichroic mirror, are used; no emission filter is needed. Cube 1 is used for CFP excitation (excitation filter, D436/20; dichroic, 455DCLP). Cube 2 is used for YFP excitation (HQ500/20; Q515LP). A spectrograph in conjunction with a cooled CCD camera is attached to the exit port of the microscope for spectroscopic imaging.

1. Just prior to fluorescence imaging, replace the culture medium for HEK 293 cells in a Nunc Lab-Tek chamber with cell imaging solution (*see* **Note 2**).
2. Put the chamber on the microscope stage. Under the transmitted light, focus on an isolated cell with a 60X Apo Fluo objective (1.45 NA).
3. Switch the light path from eyepieces to the exit port to which the spectrograph/camera compo is attached. Remove the spectrograph input slit from the light path by pushing down on the handle. Set the angle of the grating in the spectrograph for 0 nm. Take a bright field image of the cell.
4. Put the spectrograph input slit into the light path. Make sure that the slit overlays with the fluorescent cell to be imaged. Set the grating angle so that the proper range of the spectrum is projected onto the CCD camera chip.
5. Move cube 1 into the light path. Take an image, which constitutes the emission spectrum with the CFP excitation (**Fig. 2A1,A3**).
6. Draw a horizontal line across the region of the image that corresponds to the cell (**line 1** in **Fig. 2A1**). Open the linescan window in *MetaMorph*. Log the spectrum to *Excel* using the log data function (*see* **Note 3**).
7. Draw a second line across the blank region of the image (**line 2** in **Fig. 2A1**) to estimate the background level. Log the background spectrum to *Excel*.
8. Subtract the background spectrum from the first spectrum. When a CFP cell is used, this yields the CFP spectrum (**Fig. 2B1**, F_C^C); when a YFP cell is used, this yields the YFP spectrum (**Fig. 2B2**, F_C^Y); when a CFP + YFP cell is used, this yields the total spectrum (**Fig. 2B1**, F_C^{C+Y}).
9. Move cube 2 into the light path. Repeat **steps 4–7** to generate an emission spectrum with the YFP excitation. When a YFP cell or a CFP + YFP cell is used, this yields the total YFP spectrum (**Fig. 2B2,B1**, F_Y^Y).

3.1.2. Quantification of FRET Efficiency With Spectral Analysis

To quantify FRET efficiency with the spectra FRET method, spectra measurements are made from three cell groups: (1) cells expressing only CFP, from which a standard CFP spectrum (F_C^C) is collected with cube 1; (2) cells expressing only YFP, from which the spectrum collected with cube 1 represents the direct excitation of YFP with CFP excitation light (the "cross-talk" component, F_C^Y), and the spectrum collected with cube 2 represents the total YFP emission (F_Y^Y); and (3) cells expressing both CFP and YFP, from which the spectrum collected with cube 1 represents the total fluorescence emission from both CFP and YFP (F_C^{C+Y}) and the

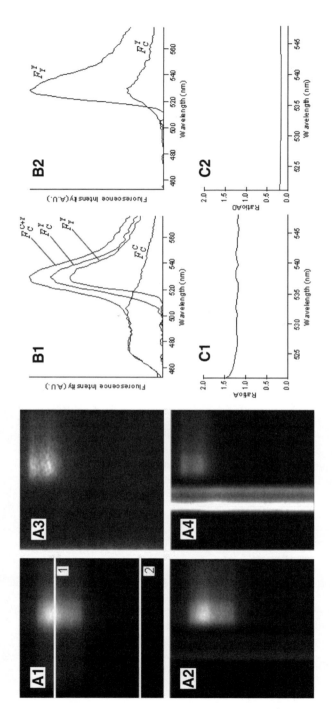

Fig. 2. Spectrum-based fluorescence resonance energy transfer quantification. (A) Spectroscopic imaging of culture cells. The emission spectra of a cell expressing cyan fluorescent protein and yellow fluorescent protein (CFP + YFP) (A1 and A2) or YFP (A3 and A4) are collected with either cube 1 (A1 and A3) or cube 2 (A2 and A4). The horizontal lines are used to measure the spectrum of the cell (line 1) and that of the background (line 2). The bright vertical bar seen in A2 and A4 is caused by excitation light that leaks through the dichroic mirror. (B) Example emission spectra measured from cells expressing CFP + YFP or only CFP (B1) or cells expressing only YFP (B2) as follows:. CFP emission spectrum (from a CFP-expressing cell);, total emission spectrum excited with cube 1 (from a CFP + YFP-expressing cell);, YFP emission spectrum excited with cube 2;, YFP emission spectrum excited with cube 1, either measured directly from a YFP expressing cell (B2) or calculated as the difference between and from a CFP + YFP-expressing cell (B1). (C) RatioA (C1) and RatioA$_0$ (C2) as functions of the wavelength.

spectrum collected with cube 2 represents the total YFP emission from these cells (F_Y^Y). After all the required spectra are measured, the apparent FRET efficiency is quantified as the enhanced emission of the acceptor (YFP) during donor (CFP) excitation.

1. Separate the CFP emission and the YFP emission from spectra of CFP + YFP cells. The standard CFP spectrum, collected from a cell expressing only CFP-tagged channels (**Fig. 2B1**, F_C^C), is scaled to the CFP peak region of the total emission spectrum from a cell expressing both CFP and YFP (**Fig. 2B1**, F_C^{C+Y}). The difference between the two spectra is calculated, which represents the YFP emission spectrum of the CFP + YFP cell. This extracted YFP spectrum (**Fig. 2B1**, F_C^Y) contains two kinds of YFP emission: the YFP emission caused by direct excitation by the CFP excitation light and the YFP emission caused by FRET.
2. The ratio of F_C^Y to the YFP spectrum with direct excitation (**Fig. 2B1**, F_Y^Y) is calculated as RatioA (*see* **Note 4**).
3. Using the two spectra from cells expressing only YFP-tagged channels, another ratio, denoted RatioA$_0$, is calculated. This is the ratio of the YFP emission spectrum collected with cube 1 (**Fig. 2B2**, F_C^Y) to the spectrum collected with cube 2 (**Fig. 2B2**, F_Y^Y). RatioA$_0$ corresponds to the cross-talk component of RatioA.
4. The apparent FRET efficiency is then calculated directly from RatioA and RatioA$_0$ as either (RatioA – RatioA$_0$) or (RatioA/RatioA$_0$). Both quantities are directly proportional to the FRET efficiency.

3.2. FRET Imaging of Oocytes Using Confocal Microscopy

The following protocol is based on a Leica TCS NT/SP confocal microscope controlled by the *LCS* software package provided by the company. A programmable sliding emission slit of the Leica microscope allows collection of the fluorescence emission at different wavelengths. Individual images of the same oocyte are collected using the identical excitation light at a series of emission wavelengths. Quantification of the fluorescence intensity of these images as a function of the emission wavelength generates an emission spectrum used to quantify FRET efficiency *(13,15,17,18)*. This approach is not limited to the Leica confocal microscope. The Zeiss LSM 510META confocal microscope, for example, is another type of confocal microscope that allows easy spectroscopic imaging, achieved by a PMT array in combination with a prism.

3.2.1. Confocal Fluorescence Imaging

1. Put an oocyte expressing fluorescent channels in a Nunc Lab-Tek chamber that contains the oocyte solution. It is preferred to have the dark side (animal pole) of the oocyte facing down (*see* **Note 5**). Focus on the lower half of the oocyte.
2. Select the proper excitation laser line (458 nm for CFP and FRET imaging, 488 nm for YFP imaging); set the laser output power (*see* **Note 6**).

3. Set the number of images to 50 and the emission window to 5 nm. Select the proper emission wavelength range (460–560 nm for CFP and FRET imaging, 490–590 nm for YFP imaging). Set the PMT gain and pinhole size (*see* **Note 6**).

4. Collect images with a ×5 objective and save them as individual TIFF files with names that end with incremental numbers (e.g., ABC001, ABC002, ABC003, and so on). This file name format greatly simplifies the following image analysis procedure described in **Subheading 3.2.2.**

5. For CFP oocytes, collect one set of images with the 458-nm excitation. For YFP oocytes and CFP + YFP coexpression oocytes, collect two sets of images, one set with the 458-nm excitation, another set with the 488-nm excitation.

3.2.2. Image Analysis

1. In *MetaMorph*, open one set of 50 images from the same oocyte by building a stack (under File menu choose Open Special …/Build Stack …/Numbered Names …; specify the first and the last images; click OK).

2. Specify two regions-of-interest (ROIs), one for the oocyte surface membrane, another one within a blank region (corresponding to the background signal).

3. Calculate the average fluorescence intensity in each ROI for each of the 50 images; build a table of the intensity values (*see* **Note 3**).

4. Subtract the background signal intensity at each wavelength from the fluorescence intensity measured from the membrane region. Plot the subtracted emission spectrum. From CFP-expressing oocytes, this yields a CFP emission spectrum (**Fig. 2B1**, F_C^C); from CFP + YFP coexpressing oocytes, it yields two spectra, one for the total YFP emission with the 488-nm excitation (**Fig. 2B1**, F_Y^Y), another for the CFP/YFP emission with the 458-nm excitation (**Fig. 2B1**, F_C^{C+Y}); from YFP-expressing oocytes, it also yields two spectra, one for the total YFP emission with the 488-nm excitation (**Fig. 2B2**, F_Y^Y), another for the cross-excitation of YFP with the 458-nm excitation (**Fig. 2B2**, F_C^Y).

5. After all the required spectra are collected from CFP, YFP, and CFP + YFP oocytes, the procedure for quantification of the FRET efficiency is identical to that described for the culture cell experiment (*see* **Subheading 3.1.2.**).

3.3. Fluorescence Intensity Ratio Analysis of Channel Subunit Stoichiometry

Ion channels are multisubunit proteins. Most native ion channels are heteromultimeric, composed of different subunit types that contribute specific properties to the channel *(27)*. Fluorescence labeling with CFP and YFP provides a new tool to study the subunit stoichiometry of intact channels under physiological conditions. Labeling of channel subunits with these fluorescent proteins often produces minor, if any, changes in channel properties (which can be tested functionally). In addition, ion channels expressed in nonnative expression systems often appear to assemble into the same heteromultimeric complex as those in native cells. Under these conditions, FRET-based studies

(17–19,21) add a complementary approach to functional studies and biochemistry studies *(28–32)* in decoding ion channel stoichiometry.

The rationale for the FRET-based channel stoichiometry analysis is that the size of a channel complex is expected to be within the range of $2R_0$ for the CFP-YFP pair (about 100 Å) *(8,9)*. Hence, the existence of FRET between different subunits tagged with CFP and YFP, respectively, would suggest that these subunits co-assemble to form heteromeric channels. In addition, the existence of FRET between the like subunits tagged with CFP and YFP, respectively, would suggest that the subunits type presents in the channel in multiple copies, and the absence of FRET would suggest that the subunit type presents in only a single copy. For many ion channels that are composed of a small number of subunits, such information from FRET experiments would be sufficient to deduce the only possible subunit stoichiometry. However, some channel types appear to be composed of a large number of subunits. For these channels, FRET experiments alone would be insufficient to deduce the definite subunit stoichiometry *(19)*.

Here, a non-FRET fluorescence method is described that, in combination with FRET, may provide sufficient information to define the subunit stoichiometry of most channel types. The idea behind this approach (which is called fluorescence intensity ratio [FIR] analysis; *[17]*) is that, because of the covalent linkage between the channel subunit and the fluorescent protein, the molar ratio between CFP and YFP molecules should be the same as the molar ratio between the subunits to which they are attached (*see* **Note 7**). The FIR approach uses the same subunit-fluorescent protein fusion constructs as well as the same instrumentation used in FRET experiments. The following protocol is designed for confocal fluorescence microscopy of oocytes with which expression levels of channel subunits can be reliably controlled by the amount of cRNA injected.

1. Divide oocytes into two groups. For group 1 oocytes, coinject each with a mixture of X-CFP and Y-YFP cRNAs, where X and Y represent different subunit types to be tested. For group 2 oocytes, coinject with a mixture of X-YFP and Y-CFP cRNAs in which the concentration ratio between X and Y is the same as that used for group 1. Incubate oocytes at 16°C for 2–7 d for channel expression.

2. From each oocyte, take a CFP image and a YFP image with a low-power (e.g., ×5) objective lens. For CFP, use 458-nm excitation and 475- to 505-nm emission; for YFP, use 488-nm excitation and 500- to 530-nm emission.

3. Measure from each oocyte the total CFP intensity F_{CFP} and the total YFP intensity (F_{YFP}) as described in **Subheading 3.2.2.** (*see* **Note 8**).

4. Make a scatterplot for each oocyte group; the CFP intensity of each oocyte is plotted against the YFP intensity of the same oocyte. If a homogeneous population of heteromeric channels is formed, then the data points should form a straight line that intersects with the origin of the coordinates. Fit the data with a linear function $F_{CFP} = K \cdot F_{YFP}$, in which k is the slope factor of the linear distribution.

5. From the slope of group 1 oocytes k_1 and that of group 2 oocytes k_2, the subunit ratio between X and Y can be calculated as $\sqrt{\frac{k_1}{k_2}}$ (*see* **Note 9**).

4. Notes

1. For the donor dequenching type of FRET quantification methods (e.g., enhanced CFP emission after YFP photobleaching), it is desirable that each CFP molecule is coupled to at least one YFP molecule. On the other hand, for FRET methods based on enhanced acceptor emission, including the spectra FRET approach described in this chapter, it is desirable that each YFP molecule is coupled to at least one CFP molecule. For the spectra FRET method, the emission from extra CFP molecules is removed when the extracted YFP spectrum is constructed.
2. The coverglass bottom of the Nunc Lab-Tek chamber allows easy fluorescence imaging. For the imaging of culture cells using an epifluorescence microscope, the no. 1 coverglass chamber is preferred to accommodate the short working distance of most high-power objectives.
3. *MetaMorph* provides a log data function for automatic tabulation of statistical values that greatly simplifies the process of building a spectrum. The following steps will set up a dynamic data exchange link between *MetaMorph* and *Excel*. Within the Linescan window (or, for the oocyte image analysis protocol described in **Subheading 3.2.2.**, choose Show Region Statistics from the Measure menu), click Open Log. Within the next popup window, check Dynamic Data Exchange (DDE), click OK. This will open another window. Set Application to Microsoft *Excel*; specify Sheet Name, Starting Row, and Starting Column; click OK. To tabulate a spectrum, click F9:Log Data (or press function key F9).
4. Because RatioA is independent of the wavelength, it can be used to conveniently check for linearity of the recording system as well as any significant contaminations by other fluorescence sources (e.g., autofluorescence).
5. Pigments on the dark side of the oocyte help prevent contamination by auto-fluorescence from the cytosolic milieu. Oocytes tend to roll over and lay on the light side (the vegetal pole). Care should be taken that the oocyte has not moved during the imaging process.
6. Make sure that the excitation and fluorescence detection parameters are set so that at the peak emission wavelengths no pixel in the image reaches saturation. All images from the same oocyte should be taken under identical condition.
7. FIR assumes that fluorescence emissions of CFP and YFP are independent. This is, of course, probably not true in most cases because of FRET between these fluorophores. FRET, which affects only measurements of the CFP intensity and not measurements of the YFP intensity, apparently has only a small effect on FIR; it takes a 50% FRET efficiency to make a 1:1 subunit ratio appear as a 1:2 ratio. If needed, the reduction of the CFP intensity caused by FRET can be corrected using the FRET efficiency measured from the same channels (*17*). Because both CFP and YFP are measured at their respective peak emissions with the direct excitation of the fluorophore, FIR is not affected by either cross-excitation or spectral bleed-through.

8. For FIR analysis, it is critical to keep the recording conditions unchanged from oocyte to oocyte. This includes the laser intensity, pinhole size, emission window size, and so on.

9. Another useful quantity, C, can be calculated as $\sqrt{k_1 \cdot k_2}$ *(17)*. This is a constant that is determined by the recording system and the fluorophores. There are two practical usages of C. First, because it is a constant the measured values for C in different experiments can be used as a reporter for erratic experimental operations, for example, accidental alteration of the laser intensity. Second, C can be used to tune the system to the optimal conditions for FIR experiments: When the system is adjusted to produce a C value close to 1, data points in an FIR plot are expected to follow the central diagonal line for a 1:1 subunit ratio and to have symmetrical distributions on both sides of the diagonal line for other ratios.

References

1. Sekar, R. B. and Periasamy, A. (2003) Fluorescence resonance energy transfer (FRET) microscopy imaging of live cell protein localizations. *J. Cell Biol.* **160,** 629–633.

2. Truong, K. and Ikura, M. (2001) The use of FRET imaging microscopy to detect protein–protein interactions and protein conformational changes in vivo. *Curr. Opin. Struct. Biol.* **11,** 573–578.

3. Hink, M. A., Bisselin, T., and Visser, A. J. (2002) Imaging protein–protein interactions in living cells. *Plant Mol. Biol.* **50,** 871–883.

4. Day, R. N., Periasamy, A., and Schaufele, F. (2001) Fluorescence resonance energy transfer microscopy of localized protein interactions in the living cell nucleus. *Methods* **25,** 4–18.

5. Stryer, L. (1978) Fluorescence energy transfer as a spectroscopic ruler. *Annu. Rev. Biochem.* **47,** 819–846.

6. Clegg, R. M. (1992) Fluorescence resonance energy transfer and nucleic acids. *Methods Enzymol.* **211,** 353–388.

7. Selvin, P. R. (1995) Fluorescence resonance energy transfer. *Methods Enzymol.* **246,** 300–334.

8. Heim, R. and Tsien, R. Y. (1996) Engineering green fluorescent protein for improved brightness, longer wavelengths and fluorescence resonance energy transfer. *Curr. Biol.* **6,** 178–182.

9. Tsien, R. Y. (1998) The green fluorescent protein. *Annu. Rev. Biochem.* **67,** 509–544.

10. Miyawaki, A. (2003) Visualization of the spatial and temporal dynamics of intracellular signaling. *Dev. Cell* **4,** 295–305.

11. Butkevich, E., Hulsmann, S., Wenzel, D., Shirao, T., Duden, R., and Majoul, I. (2004) Drebrin is a novel connexin-43 binding partner that links gap junctions to the submembrane cytoskeleton. *Curr. Biol.* **14,** 650–658.

12. Mori, M. X., Erickson, M. G., and Yue, D. T. (2004) Functional stoichiometry and local enrichment of calmodulin interacting with Ca^{2+} channels. *Science* **304,** 432–435.

13. Trudeau, M. C. and Zagotta, W. N. (2004) Dynamics of Ca^{2+}-calmodulin-dependent inhibition of rod cyclic nucleotide-gated channels measured by patch-clamp fluorometry. *J. Gen. Physiol.* **124,** 211–223.

14. Erickson, M. G., Alseikhan, B. A., Peterson, B. Z., and Yue, D. T. (2001) Preassociation of calmodulin with voltage-gated Ca(2+) channels revealed by FRET in single living cells. *Neuron* **31,** 973–985.

15. Zheng, J., Varnum, M. D., and Zagotta, W. N. (2003) Disruption of an intersubunit interaction underlies Ca^{2+}-calmodulin modulation of cyclic nucleotide-gated channels. *J. Neurosci.* **23,** 8167–8175.

16. Tsuboi, T., Lippiat, J. D., Ashcroft, F. M., and Rutter, G. A. (2004) ATP-dependent interaction of the cytosolic domains of the inwardly rectifying K^+ channel Kir6.2 revealed by fluorescence resonance energy transfer. *Proc. Natl. Acad. Sci. USA* **101,** 76–81.

17. Zheng, J. and Zagotta, W. N. (2004) Stoichiometry and assembly of olfactory cyclic nucleotide-gated channels. *Neuron* **42,** 411–421.

18. Zheng, J., Trudeau, M. C., and Zagotta, W. N. (2002) Rod cyclic nucleotide-gated channels have a stoichiometry of three CNGA1 subunits and one CNGB1 subunit. *Neuron* **36,** 891–896.

19. Staruschenko, A., Medina, J. L., Patel, P., Shapiro, M. S., Booth, R. E., and Stockand, J. D. (2004) Fluorescence resonance energy transfer analysis of subunit stoichiometry of the epithelial Na^+ channel. *J. Biol. Chem.* **279,** 27,729–27,734.

20. Biskup, C., Zimmer, T., and Benndorf, K. (2004) FRET between cardiac Na^+ channel subunits measured with a confocal microscope and a streak camera. *Nat. Biotechnol.* **22,** 220–224.

21. Amiri, H., Schultz, G., and Schaefer, M. (2003) FRET-based analysis of TRPC subunit stoichiometry. *Cell Calcium* **33,** 463–470.

22. Schaefer, M., Plant, T. D., Stresow, N., Albrecht, N., and Schultz, G. (2002) Functional differences between TRPC4 splice variants. *J. Biol. Chem.* **277,** 3752–3759.

23. Hosaka, Y., Hanawa, H., Washizuka, T., et al. (2003) Function, subcellular localization and assembly of a novel mutation of KCNJ2 in Andersen's syndrome. *J. Mol. Cell Cardiol.* **35,** 409–415.

24. Nashmi, R., Dickinson, M. E., McKinney, S., et al. (2003) Assembly of alpha4beta2 nicotinic acetylcholine receptors assessed with functional fluorescently labeled subunits: effects of localization, trafficking, and nicotine-induced upregulation in clonal mammalian cells and in cultured midbrain neurons. *J. Neurosci.* **23,** 11,554–11,567.

25. Gordon, G. W., Berry, G., Liang, X. H., Levine, B., and Herman, B. (1998) Quantitative fluorescence resonance energy transfer measurements using fluorescence microscopy. *Biophys. J.* **74,** 2702–2713.

26. Xia, Z. and Liu, Y. (2001) Reliable and global measurement of fluorescence resonance energy transfer using fluorescence microscopes. *Biophys. J.* **81,** 2395–2402.

27. Hille, B., (2001) *Ion Channels of Excitable Membranes*, 3rd ed. Sinauer, Sunderland, MA.

28. Weitz, D., Ficek, N., Kremmer, E., Bauer, P. J., and Kaupp, U. B. (2002) Subunit stoichiometry of the CNG channel of rod photoreceptors. *Neuron* **36,** 881–889.
29. Zhong, H., Molday, L. L., Molday, R. S., and Yau, K. W. (2002) The heteromeric cyclic nucleotide-gated channel adopts a 3A:1B stoichiometry. *Nature* **420,** 193–198.
30. Firsov, D., Gautschi, I., Merillat, A. M., Rossier, B. C., and Schild, L. (1998) The heterotetrameric architecture of the epithelial sodium channel (ENaC) *EMBO J.* **17,** 344–352.
31. Snyder, P. M., Cheng, C., Prince, L. S., Rogers, J. C., and Welsh, M. J. (1998) Electrophysiological and biochemical evidence that DEG/ENaC cation channels are composed of nine subunits. *J. Biol. Chem.* **273,** 681–684.
32. Eskandari, S., Snyder, P. M., Kreman, M., Zampighi, G. A., Welsh, M. J., and Wright, E. M. (1999) Number of subunits comprising the epithelial sodium channel. *J. Biol. Chem.* **274,** 27,281–27,286.

III

METHODS FOR STUDYING CHANNEL REGULATION AND PHYSIOLOGICAL FUNCTION

7

Probing the Effects of Phosphoinositides on Ion Channels

Chou-Long Huang

Summary

Ion channels are integral membrane proteins that control transmembrane ion fluxes to regulate membrane potential, cell excitability, and ion transport. Membrane phospholipids containing phosphoinositides have recently emerged as important regulators of many ion channels, including inward rectifier K^+ channel, voltage-gated K^+ and Ca^{2+} channels, transient receptor potential channels, and intracellular inositol-1,4,5-trisphosphate receptor ion channels. Discussed here are several methods for studying regulation of ion channels by phospholipids.

Key Words: Anti-PIP_2 antibody; patch-clamp; phosphoinositides; phospholipase C; PIP_2; polylysine; two-electrode voltage-clamp; wortmannin; *Xenopus* oocytes.

1. Introduction

There are multiple biologically active phosphoinositides in the membrane that regulate diverse eukaryotic cellular processes, including vesicular transport, growth factor and calcium signaling, organization of the cytoskeleton, and ion channel and transporter activity (*1,2*). In addition to differences in target specificity, the concentration of phosphoinositide isoforms in the cell membrane varies considerably (*2*). Phosphatidylinositol 4,5-bisphosphate [PI(4,5)$_2$], the precursor of inositol-1,4,5-trisphosphate (IP$_3$) and diacylglycerol, is the most abundant phosphatidylinositol bisphosphate (*2–4*). It comprises approx 1% of the total pool of phospholipids in the plasma membrane. The cellular concentration of PI(4,5)$_2$ (normalized to the total cell volume) is approx 10 μM (*3*). The effective concentration in the membrane (normalized to the inner leaflet of the plasma membrane) is, however, much higher and estimated at approx 5 mM (*3*). The membrane concentration of phosphatidylinositol-4-phosphate is approx equal to that of the PI(4,5)P$_2$. In contrast, the D-3 isoforms, phosphati-

From: *Methods in Molecular Biology, vol. 337: Ion Channels: Methods and Protocols*
Edited by: J. D. Stockand and M. S. Shapiro © Humana Press Inc., Totowa, NJ

dylinositol-3,4-bisphosphate [PI(3,4)P$_2$] and phosphatidylinositol-3,4,5-trisphosphate [PI(3,4,5)P$_3$] are considerably less abundant, with membrane concentration approx 10 μM for both. The membrane concentration of phosphatidylinositol-3-phosphate is approx 200 μM. The precursor of all phosphoinositides, phosphatidylinositol, is abundantly present in the membrane, with membrane concentration approx 100 mM.

Several experimental approaches are available for altering the content of phosphoinositides in the membrane for studying regulation of ion channels by the lipids. These include direct application of lipids or lipid inhibitors to excised inside-out membranes, intracellular delivery of lipids using injection or via whole-cell patch pipets or lipid carriers, and alteration of lipids *in situ* using physiological ligands.

2. Materials

1. Purified bovine brain PI(4,5)P$_2$ and phosphatidylinositol-4-phosphate are available from Roche (Indianapolis, IN). Other suppliers include Calbiochem (La Jolla, CA) and Avanti Polar Lipids (Alabaster, AL). PI(3,4)P$_2$ and PI(3,4,5)P$_3$ are much less abundant in biological membranes. Synthetic versions can be obtained from Avanti Lipids, Echelon (Salt Lake City, UT), and Matreya (Pleasant Gap, PA). Short-chain synthetic dioctanoyl (DiC$_8$) lipids are also available from these suppliers. Phosphoinositides can be prepared by dissolving 1 mg powder in water (1 mM) and sonicating in a water bath sonicator for 20 s three times. Avoid heat production during sonication. Store lipids in 20-μL aliquots at −20 or −70°C.
2. Anti-PIP$_2$ antibody is available from Assay Designs (Ann Arbor, MI). Dissolve protein A-purified antibody (100 μg) in water (100 μL). Store in small aliquots at −20 or −70°C.
3. Polylysine of various average molecular weights (1–7 kDa) is available from Sigma (St. Louis, MO). We have used polylysine with an average molecular weight of 1 kDa at 30 μg/mL. Others have used polylysine with an average molecular weight of 7 kDa at 300 μg/mL.
4. Heparin is available from Sigma. Effective final concentration ranges from 300 to 1000 μg/mL.
5. Type III-S histone from calf thymus is available from Sigma.
6. Thrombin (human α-thrombin) is available from Enzyme Research Laboratories (South Bend, IN).
7. Wortmannin is available from Sigma. Prepare wortmannin stocks by dissolving in dimethyl sulfoxide (10 mM) and store aliquots at −20°C.
8. Fluoride vanadate pyrophosphate (FVPP) bathing solution for patch clamp studies: 100 mM KCl, 10 mM HEPES at pH 7.4, 5 mM ethylenediaminetetraacetic acid, 5 mM NaF, 0.1 mM Na$_3$VO$_4$, and 10 mM Na$_3$HP$_2$O$_7$ (*5,6*).
9. Bath solution containing Mg^{2+}: 100 mM KCl, 10 mM HEPES at pH 7.4, 5 mM EGTA, and 1 mM MgCl$_2$ (*5,6*).

10. Mg-ATP (adenosine triphosphate) solution used for regenerating lipids via endogenous lipid kinases *in situ*: 100 mM KCl, 10 mM HEPES at pH 7.4, 5 mM EGTA, and 1 mM Mg-ATP *(5,6)*.

11. Whole-cell intracellular pipet solution containing Mg-ATP and Mg-GTP (guanosine 5′-triphosphate) used to study phospholipid regulation of KCNQ2/KICNG3 K$^+$ channels: 175 mM KCl, 5 mM MgCl$_2$, 5 mM HEPES at pH 7.4, 0.1 mM BAPTA, 3 mM Na$_2$ATP, and 0.1 mM Na$_3$GTP.

3. Methods

3.1. Application of Phosphoinositides and Inhibitors to Excised Inside-Out Patches

3.1.1. Application of Phosphoinositides

1. Perform cell-attached patch clamp recording. Excise inside-out patches into a Mg^{2+}-free solution containing a mixture of phosphatase inhibitors to form excised inside-out patches. For inward rectifier K$^+$ channels, we use the FVPP bath solution (containing the phosphatase inhibitors fluoride vanadate pyrophosphate; *[5,6]*). Methods of cell-attached and inside-out patch clamp recording of native or expressed channels can be reviewed in **ref. 7**.

2. After currents stabilize in inside-out patches, change cytoplasmic solution to a Mg^{2+}-containing solution. Allow channels to run down (decrease in activity) in this Mg^{2+}-containing solution (*see* **Note 1**).

3. After channels run down, return bathing solution back to FVPP.

4. Thaw lipid aliquots, mix vigorously in a vortex mixer, and keep on ice until use. Dilute lipids in FVPP solution just prior to use.

5. Perfuse lipids in FVPP solution to the (bath) cytoplasmic face of the channel in inside-out patches. We have used manual perfusion (using a 1-mL syringe) at approx 15 μL/s for 20–30 s. Within 20–30 s, 1–10 μM PI(4,5)P$_2$ reactivates inward rectifier K$^+$ channels from rundown *(5,6,8)*.

6. Efficacy and specificities of different phosphoinositides can be tested using inside-out patches (*see* **Note 2**).

3.1.2. Application of Anti-PIP$_2$ Antibody

1. Perform cell-attached patch clamp recording. Excise membrane patches into FVPP solution. FVPP solution contains inhibitors of lipid phosphatases, which prevent loss of phosphoinositides on excision.

2. Dilute anti-PIP$_2$ antibody 50-fold in FVPP solution. After current stabilizes in FVPP solution, apply anti-PIP$_2$ antibody to the cytoplasmic face of the inside-out membrane. If a channel is activated by PI(4,5)P$_2$, then application of antibody decreases activity (*see* **Note 3**).

3. Reverse the effects of anti-PIP$_2$ antibody by applying 2 mM dithiothreitol. Dithiothreitol inactivates the anti-PIP$_2$ antibody by reducing disulfide bonds within this antibody.

3.1.3. Application of Polylysine

1. Perform cell-attached patch clamp recording. Excise patches into FVPP cytoplasmic solution.
2. Dilute polylysine in FVPP. After current stabilizes in FVPP solution, apply polylysine to the cytoplasmic face of the inside-out membrane.
3. Apply heparin to the inside-out patch to reverse the effect of polylysine.

3.1.4. Application of Mg-ATP to Regenerate Lipids via Lipid Kinases In Situ

1. Perform cell-attached patch clamp recording. Excise patches into FVPP cytoplasmic solution.
2. After currents stabilize, change cytoplasmic solution to a Mg^{2+}-containing solution. Allow channels to run down in the Mg^{2+}-containing solution.
3. Apply a cytoplasmic solution containing Mg-ATP to reactivate the channels. Mg-ATP stimulates lipid kinases to generate lipid *in situ*.
4. Confirm regeneration of $PI(4,5)P_2$ by applying anti-PIP_2 antibody to antagonize reactivation (*see* **Note 4**).

3.2. Intracellular Delivery of Phosphoinositides in Cell-Attached and Whole-Cell Recording

3.2.1. Intracellular Delivery of Phosphoinositides in Cell-Attached Recordings Using Histone as a Lipid Carrier

1. Perform cell-attached patch clamp recording.
2. Thaw one aliquot of lipid, mix vigorously in a vortex mixer, and keep on ice until use. Just prior to use, dilute equal molar lipid and histone in the extracellular recording solution and mix in a vortex mixer. For example, 10 µM $PI(4,5)P_2$ and 10 µM histone (*see* **Note 5**).
3. Exchange the extracellular solution with one that contains lipid–histone complex. Changes in ion channel activity in response to lipid delivery via histone carriers may be expected in a few minutes. Control experiments will include application of $PI(4,5)P_2$ or histone alone to the extracellular solution.

3.2.2. Intracellular Delivery of Phosphoinositides Via Patch Pipets in Whole-Cell Recordings

1. Dilute lipids in intracellular pipet solution to 10–20 µM.
2. Perform ruptured whole-cell recording using Mg^{2+}-free pipet solution containing lipids.

3.2.3. Intracellular Injection of Phosphoinositides in Whole-Cell Recordings of Xenopus Oocytes

1. Thaw an aliquot of lipid stock (1 mM) and mix vigorously by vortexing.
2. Inject oocytes with lipid stock or water.
3. Perform whole-cell recording using two-electrode voltage clamp on lipid- or water-injected *Xenopus* oocytes (*9*).

3.3. Decrease PI(4,5)P₂ Content In Situ *Via Phospholipase C-Coupled Hormone Receptors*

1. Express recombinant channels of interest in Chinese hamster ovary cells. Chinese hamster ovary cells contain endogenous thrombin receptors *(10)*.
2. Perform whole-cell patch clamp recording using an intracellular pipet solution containing Mg-ATP and Mg-GTP (*see* **ref. *11*** for investigation of PIP₂ regulation of KCNQ2/KCNQ3 K⁺ channels).
3. Add wortmannin (final concentration 10 μM) to the extracellular solution and incubate for 5 min to inhibit phosphatidylinositol 4-kinase (*see* **Note 6**).
4. Add thrombin (final concentration 2 U/mL) to the extracellular solution to stimulate phospholipase C activity via thrombin receptors (*see* **Notes 7** and **8**).

4. Notes

1. The first clue that a channel is regulated by phosphoinositides usually comes from the observation that channel activity changes on excision of inside-out patches into Mg^{2+}-containing, ATP-free solution. In a resting state, membrane phosphoinositides are maintained by dynamic synthesis and degradation via kinases and phosphatases, respectively. Exposing the cytoplasmic face of membranes to a Mg^{2+}-containing, ATP-free solution activates lipid phosphatases preferentially to reduce phosphoinositides in the membrane. Reduction of channel activity in the inside-out patch (rundown) may be caused by other factors in addition to loss of phosphoinositides. Reactivation of a channel by phospholipids following rundown confirms and identifies which phosphoinositide regulates the channel.
2. Different phosphoinositides can be applied to inside-out patches to examine their efficacy and specificities. Phosphoinositides in the native membrane contains a mixture of different acyl chains. The most abundant form contains arachidonyl and stearyl at positions 2 and 3, respectively. PI(3,4)P₂ and PI(3,4,5)P₃ are much less abundant in biological membranes. To avoid differences because of the acyl chains, we and other investigators routinely use standardized, synthetic lipids, such as dipalmitoyl (DiC₁₆) and DiC₈ analogs *(8,12)*. The DiC₈ analogs have a higher aqueous-lipid partition coefficient. Thus, the effects of the DiC₈ analogs are reversible by extensive washout *(8,12)*.
3. For inward rectifier K⁺ channel, direct interaction between anionic phosphoinositides and basic amino acids in the intracellular region of the channels stabilizes open conformation. Anti-PIP₂ antibody sequesters PI(4,5)P₂ to prevent activation of channels *(5,6,8)*. The time required for 50% inhibition by anti-PIP₂ antibody (expressed as half-time for inhibition, $T_{1/2}$) correlates well with the affinity of channels for PIP₂ *(5,9,13)*.
4. Application of Mg-ATP in the excised patches reactivates inward rectifier K⁺ channels after rundown *(14)*. Besides activation of lipid kinases to regenerate PI(4,5)P₂ in the membrane, Mg-ATP may activate protein kinases. Inhibition by anti-PIP₂ antibody supports the idea that Mg-ATP stimulates lipid kinases to pro-

duce PI(4,5)P$_2$, with the phospholipid then activating the channel *(6)*. In addition, adenosine-5′-[γ-thio]triphosphate serves as a substrate for some protein kinases (such as protein kinase A) but not for lipid kinases. Mg-adenosine-5′-[γ-thio]triphosphate can be used to distinguish the effects of protein vs lipid kinases *(6)*.

5. Other polybasic compounds, including neomycin and polyamine, have also been used as carriers to facilitate uptake of phosphoinositides by cells *(15)*. However, histone is the least toxic to most cells.

6. Wortmannin inhibits phosphatidylinositol 3-kinase with an IC$_{50}$ of approx 5 n*M*. It also inhibits phosphatidylinositol 4-kinase, although with a higher IC$_{50}$ of approx 100 n*M* *(16)*. At 10 µ*M*, wortmannin inhibits phosphatidylinositol 4-kinase and prevents resynthesis of PI(4,5)P$_2$ following hydrolysis by phospholipase C *(16)*. An additional way to prevent resynthesis of PI(4,5)P$_2$ is to use an intracellular pipet solution lacking Mg-ATP *(11)*.

7. The abundance of PI(3,4,5)P$_3$ and PI(3,4)P$_2$ in the plasma membrane is approx 1% of PI(4,5)P$_2$ but may rise to approx 10% when phosphatidylinositol 3-kinase is stimulated *(3)*. A physiologically important regulation by PI(3,4,5)P$_3$ and PI(3,4)P$_2$ will require that PI(3,4,5)P$_3$ and PI(3,4)P$_2$ have a 10-fold higher specificity for the target than PI(4,5)P$_2$. Some protein domains bind PI(3,4,5)P$_3$ and PI(3,4)P$_2$ with such specificity *(2)*. Whether the PI(3,4,5)P$_3$- and PI(3,4)P$_2$-rich lipid domains exist remains unknown. Several ion channels are activated by PI(3,4,5)P$_3$ and PI(3,4)P$_2$ in excised patches *(12)*. However, a physiological role of this regulation awaits further studies.

8. A very important role of phosphoinositides is to regulate trafficking of membrane proteins, including ion channels *(2,17)*. Phosphoinositides may stimulate retrieval or insertion of ion channel-containing vesicles even in excised patches. Identification of phosphoinositide interaction domains on ion channels will help to establish direct activation by the lipids.

Acknowledgments

Work in my laboratory is supported by grants from the National Institutes of Health (DK54368, DK59530, DK20543) and an Established Investigator Award from American Heart Association (0440019N).

References

1. Hilgemann, D. W., Feng, S., and Nasuhoglu, C. (2001) The complex and intriguing lives of PIP$_2$ with ion channels and transporters. *Sci. STKE.* RE19-25.
2. Overduin, M., Cheever, M. L., and Kutateladze, T. G. (2001) Signaling with phosphoinositides: better than binary. *Mol. Interventions* **1**, 150–159.
3. Stephens, L. R., Jackson, T. R., and Hawkins, P. T. (1993) Agonist-stimulated synthesis of phosphatidylinositol (3,4,5)-trisphosphate: a new intracellular signalling system. *Biochim. et Biophys. Acta* **1179**, 27–75.
4. McLaughlin, S., Wang, J., Gambhir, A., and Murray, D (2002) PIP$_2$ and proteins: interactions, organization, and information flow. *Annu. Rev. Biophys. Biomol. Struct.* **31**, 151–175.

5. Huang, C.-L., Feng, S., and Hilgemann, D. W. (1998) Direct activation of inward rectifier potassium channels by PIP$_2$ and its stabilization by G$_{\beta\gamma}$. *Nature* **391,** 803–806.

6. Liou, H.-H., Zhou, S. S., and Huang, C.-L. (1999) Phosphorylation of ROMK1 channel by PKA regulates channel activity via a PIP$_2$-dependent mechanism. *Proc. Natl. Acad. Sci. USA* **96,** 5820–5825.

7. Rudy, B. and Iverson, L. E. (1993) *Methods in Enzymology, Vol. 207, Ion Channels,* Academic, San Diego.

8. Zeng, W.-Z., Liou, H.-H., Krishna, U. M., Falck, J. R., and Huang, C.-L. (2002) Structural determinants and specificities for ROMK1-phosphoinositide interaction. *Am. J. Physiol.* **282,** F826–F834.

9. Baukrowitz, T., Schulte, U., Oliver, D., et al. (1998). PIP$_2$ and PIP as determinants for ATP inhibition of K$_{ATP}$ channels. *Science* **282,** 1141–1144.

10. Dickenson, J. M. and Hill, S. J. (1997). Transfected adenosine A1 receptor-mediated modulation of thrombin-stimulated phospholipase C and phospholipase A2 activity in CHO cells. *Eur. J. Pharmacol.* **321,** 77–86.

11. Suh, B.-C. and Hille, B. (2002) Recovery from muscarinic modulation of M current channel requires phosphatidylinositol 4,5-bisphosphate. *Neuron* **35,** 507–520.

12. Rohacs, T., Chen, J., Prestwich, G. D., and Logothetis, D. E. (1999) Distinct specificities of inwardly rectifying K$^+$ channels for phosphoinositides. *J. Biol. Chem.* **274,** 36,065–36,072.

13. Zhang, H., He, C., Yan, X., Mirshahi, T., and Logothetis, D. E. (1999) Activation of inwardly rectifying K$^+$ channels by distinct PtdIns(4,5)P$_2$ interactions. *Nat. Cell Biol.* **1,** 183–188.

14. Hilgemann, D. W. and Ball, R. (1996) Regulation of cardiac Na$^+$, Ca^{2+} exchanges and K$_{ATP}$ potassium channels by PIP$_2$. *Science* **273,** 956–959.

15. Ozaki, S., DeWald D. B., Shope, J. C., Chen, J., and Prestwich, G. D. (2000) Intracellular delivery of phosphoinositides and inositol phosphates using polyamine carriers. *Proc. Natl. Acad. Sci. USA* **97,** 11,286–11,291.

16. Nakanishi, S., Catt, K. J., and Balla, T. (1995) A wortmannin-sensitive phosphatidylinositol 4-kinase that regulates hormone-sensitive pools of inositol-phospholipids. *Proc. Natl. Acad. Sci. USA* **92,** 5317–5321.

17. Bezzerides, V. J., Ramsey, I. S., Kotecha, S., Greka, A., and Clapham, D. E. (2004) Rapid vesicular translocation and insertion of TRP channels. *Nat. Cell Biol.* **6,** 709–720.

8

Epithelial Sodium Channel in Planar Lipid Bilayers

Bakhrom K. Berdiev and Dale J. Benos

Summary

Amiloride-sensitive Na$^+$ channels belong to the epithelial Na$^+$ channel (ENaC)-degenerin superfamily of ion channels. In addition to their key role in sodium handling, they serve diverse functions in many tissues. Improper functioning of ENaC has been implicated in several diseases, including salt-sensitive hypertension (Liddle's syndrome), salt-wasting syndrome (pseudohypoaldosteronism type I), pulmonary edema, and cystic fibrosis. We have utilized planar lipid bilayers, a well-defined system that allows simultaneous control of "internal" and "external" solutions, to study ENaCs.

Key Words: Amiloride; bilayers; channel; function; lipids.

1. Introduction

Since its introduction in the early 1960s *(1,2)*, planar lipid bilayers not only established ion channels as one of a broad class of ion-transporting pathways, but also provided first direct observation of ion currents through single ion channels (for a review, *see* **ref. 3**). Bilayers also extended our understanding of the epithelial Na$^+$ channel (ENaC) role in epithelial Na$^+$ transport. Following the cloning of the three ENaC *(4–6)* subunits, we used planar lipid bilayers to study ENaC *(7,8)*. Here, we briefly summarize two methods of ENaC protein preparation for incorporation into planar lipid bilayers.

2. Materials
2.1. In Vitro Transcription and Translation of ENaC Proteins and Reconstitution in Proteoliposomes

1. Plasmid containing full-length ENaC cDNA following the T7 transcription promoter (*see* **Note 1**).
2. Restriction endonucleases used to linearize plasmids (e.g., *Not*I; *see* **Note 2**).

From: *Methods in Molecular Biology, vol. 337: Ion Channels: Methods and Protocols*
Edited by: J. D. Stockand and M. S. Shapiro © Humana Press Inc., Totowa, NJ

3. Plasmid purification reagents (e.g., GeneClean Kit, Bio101, La Jolla, CA; *see* **Note 3**).
4. In vitro transcription reagents (e.g., Ribomax Kit, Promega, Madison, WI; *see* **Note 3**).
5. RNA cap analog $m^7G(5')ppp(5')G$ (NEB, Beverly, MA).
6. Guanosine 5'-triphosphate.
7. In vitro translation reagents (e.g., TNT transcription-translation kit; Promega; *see* **Note 3**).
8. Canine microsomes (Promega).
9. Radiolabeled methionine (e.g., [^{35}S]Trans label, ICN, Costa Mesa, CA).
10. Phosphatidylethanolamine, phosphatidylserine, and phosphatidylcholine (Avanti Polar Lipids, Alabaster, AL; *see* **Note 3**).
11. Triton X-100.
12. Liposome preparation buffer: 60 mM Tris-HCl pH 6.8 and 25% glycerol (v/v).
13. Sodium dodecyl sulfate-free sample buffer: 60 mM Tris-HCl pH 6.8, 25% glycerol, and 0.01% bromophenol blue.
14. G-150 superfine Sephadex gel filtration column (Pharmacia Biotech).
15. Gel filtration elution buffer: 500 mM NaCl, 1 mM ethylenediaminetetraacetic acid, 10 mM Tris-HCl pH 7.6.
16. Reaction buffer: 400 mM KCl, 5 mM Tris-HCl pH 7.4, and 0.5 mM $MgCl_2$.
17. Bio-Beads SM-2 (Bio-Rad, Melville, NY).
18. Dithiothreitol (DTT).

2.2. Preparation and Injection of Xenopus laevis Oocytes With ENaC cRNA and Preparation of Oocyte Plasma Membrane Vesicles Containing ENaC

1. General anesthesia for amphibians.
2. Ca^{2+}-free OR-2 solution: 82.5 mM NaCl, 2.4 mM KCl, 1.8 mM $MgCl_2$, and 5.0 mM HEPES at pH 7.4.
3. Collagenase type 1A.
4. Leibovitz L-15 buffer (Sigma, St. Louis, MO).
5. ENaC cRNA in nuclease-free water (*see* **Note 4**).
6. High-K^+ oocytes rinse buffer: 400 mM KCl, 5 mM piperazine-N,N-*bis*(2-ethansulfonic acid) pH 6.8.
7. Protease inihibitors: phenylmethylsulfonyl fluoride, pepstatin, aprotinin, and leupeptin.
8. DNase (deoxyribonuclease) I.
9. Sucrose.
10. Pellet resuspension buffer: 100 mM KCl and 5 mM MOPS pH 6.8.

2.3. Planar Lipid Bilayers

1. Bilayer system (Warner Instruments, Inc., Hamden, CT; *see* **Note 5**).
2. Ag/AgCl electrodes embedded in a 3 M KCl/3% agar bridges (*see* **Note 6**).

3. *Cis* and *trans* bathing solutions: 100 m*M* NaCl pH 7.4 (*see* **Note 7**).
4. 10 mg/mL diphytanoyl-phosphatidyl-ethanolamine (Avanti Polar Lipids) stock in chloroform.
5. 10 mg/mL diphytanoyl-phosphatidyl-serine (Avanti Polar Lipids) stock in chloroform.
6. Nitrogen gas.
7. *n*-Decane or *n*-octane.

3. Methods
3.1. In Vitro Transcription and Translation of ENaC Proteins (7–9)

1. Linearize ENaC plasmids (pSport) overnight with *Not*I and purify linearized DNA using GeneClean Kit.
2. Perform in vitro transcription using T7 RNA polymerase according to manufacturer's instructions (Ribomax Kit). Add a cap analog, m^{7}G(5')ppp(5')G in a 2:1 molar ratio to guanosine 5'-triphosphate. This results in the synthesis of a 5'-capped transcript that stabilizes the messenger RNA and increases its translation efficiency *(7)*.
3. In vitro translate individual ENaC proteins using the TNT transcription-translation kit according to the manufacturer's instructions in the presence of canine microsomal membranes and 0.8 mCi/mL [^{35}S]Trans label.
4. Mix 25 µL translation reaction for each cRNA with 0.5 mg phosphatidylethanolamine, 0.3 mg phosphatidylserine, 0.2 mg phosphatidylcholine, Triton X-100 (0.2% v/v final concentration), and 25 µL liposome preparation buffer.
5. Briefly vortex the mixture and incubate at room temperature for 5 min.
6. Add an equal volume of sodium dodecyl sulfate-free sample buffer and overlay mixture on G-150 superfine Sephadex gel filtration column.
7. Elute from column with gel filtration elution buffer.
8. Collect fractions (100 µL) and perform autoradiography to identify fractions containing ENaC proteins.
9. To prepare control liposomes, follow the same procedure except omit ENaC RNA from the reaction mixture.

3.2. Reconstitution of ENaC Proteins Into Proteoliposomes (7,8)

1. Incubate 5–10 µL of eluted ENaC proteins with 0.5 mg phosphatidylethanolamine, 0.3 mg phosphatidylserine, and 0.2 mg phosphatidylcholine.
2. Bring volume to 600 µL with reaction buffer.
3. Incubate reaction with 150 mg Bio-Beads SM-2 to remove Triton X-100.
4. Place samples in a rotator for 45 min at room temperature and then incubate overnight at 4°C.
5. Separate proteoliposomes from beads (beads will pellet overnight) using a 1-mL syringe.
6. Sonicate proteoliposomes (43 kHz, 160 W, 40 s) in the presence of 25 µ*M* dithiothreitol (DTT) and allow to reform by freeze-thawing three to five times. This

protocol helps to dissociate putative individual conduction elements of ENaC held together by sulfhydryl bonds *(8)*. With this procedure, single channels with uniform conductance of 13 pS could be observed in over 70% of total incorporations. If multiple-channel incorporation occurs, then the sonication-freeze/thawing procedure should be repeated.

7. Divide proteoliposomes into 25-µL aliquots and store at –80°C.

3.3. Preparation and Injection of Xenopus *Oocytes (7,8)*

1. From an anesthetized adult female *X. laevis* kept at 18°C in chlorine-free water, surgically remove oocytes.
2. Perform oocyte defolliculation for 2 h with Ca^{2+}-free OR-2 solution supplemented with collagenase type 1A (1 mg/mL). Solution should be exchanged at least once after 1 h.
3. After choosing and isolating stage V/VI oocytes, allow them to recover overnight in half-strength Leibovitz L15 buffer.
4. Inject oocytes with 2 ng/subunit cRNA (in 50 nL nuclease-free water) for each ENaC subunit (ENaC is comprised of three subunits, giving a total of 6 ng cRNA; *see* **Note 8**). Use as a control oocytes injected with 50 nL water only (no RNA).

3.4. Preparation of Xenopus *Oocytes Membrane Vesicles (10)*

1. Rinse 30–40 injected oocytes with high-K^+ buffer supplemented with 100 µ*M* phenylmethylsulfonyl fluoride, 1 µ*M* pepstatin, 1 µg/mL aprotinin, 1 µg/mL leupeptin, 1 µg/mL DNase I, and 300 m*M* sucrose.
2. Homogenize oocytes in 300 µL (~10 µL/oocyte) of the same buffer for 5 min using a ground glass tissue grinder.
3. Layer homogenate on a discontinuous sucrose gradient (3 mL 50% on bottom and 3 mL 20% on the top in high-K^+ buffer in the presence of protease inhibitors).
4. Centrifuge at 23,500*g* for 30 min.
5. Remove the top layer and collect the interface (white cloudy layer).
6. Dilute threefold with high-K^+ buffer.
7. Centrifuge again at 23,500*g* for 30 min.
8. Discard the supernatant and resuspend pellet in 100 µL of resuspension buffer. Divide into approx 15-µL aliquots and freeze ENaC-containing vesicles at –80°C.

3.5. Planar Lipid Bilayer System (7,8)

1. Bilayer chambers include two aqueous compartments (designated *cis* and *trans*) separated by a polycarbonate or Teflon septum containing a small (usually 150–200 µm) hole onto which the planar membrane is formed. The *cis* compartment is connected to the voltage source via an Ag/AgCl electrode and a 3 *M* KCl/3% agar bridge. The current-to-voltage converter is connected to the *trans* side of the bilayer chamber using an Ag/AgCl electrode and another 3 *M* KCl/3% agar bridge and serves as a virtual ground (*see* **Note 9**).
2. A mixture of negatively charged (phosphatidylserine) and neutral (phosphatidylethanolamine) lipids is recommended to increase the chances of fusion. The

ratio can be varied in some cases (*see* **Note 10**). Lipids are commercially available from Avanti Polar Lipids (Alabaster, AL) and should be stored at −20°C. The bilayer-forming solution (*see* **step 3**) should be made daily.

3. Bilayer formation: The "painting" technique *(1,2)* and "folded" bilayers *(11)* are the two main approaches for bilayer formation. For the painting approach, a bilayer is formed by applying a small amount of lipid solution over the septum aperture separating the two aqueous compartments. The folded bilayers are assembled at an air–water interface. Two monolayers are spread in each compartment of a bilayer chamber separated by the aperture. After allowing time for the solvent to evaporate, the buffer in each compartment is raised above the aperture, one side at a time. This procedure results in the formation of a flat lipid bilayer. A criterion for good bilayer formation is membrane capacitance of 200–300 pF (0.67–0.95 µF/cm^2).

3.6. ENaC Incorporation Into Bilayers

1. Combine 16.66 µL stock diphytanoyl-phosphatidyl-ethanolamine with 8.33 µL stock diphytanoyl-phosphatidyl-serine in a glass vial.
2. Dry under flowing nitrogen.
3. Dissolve dried lipids in 25–50 µL *n*-decane (or *n*-octane) to achieve a final concentration of 12.5–25 mg/mL.
4. Fill bilayer *cis* and *trans* chambers with bathing solution.
5. Connect the current–voltage converter to the *trans* chamber of the bilayer system using a Ag/AgCl electrode and 3 *M* KCl-3% agar.
6. Connect the voltage source to the *cis* chamber of the bilayer system using a Ag/AgCl electrode and 3 *M* KCl-3% agar.
7. Form a bilayer over the septum aperture using the lipid-containing membrane-forming solution in *n*-decane described in **step 3** (*see* **Note 11**).
8. Following formation of the bilayer membrane (ascertained by following membrane capacitance; *see* **step 7**), place a small aliquot of the ENaC-containing oocyte vesicle suspension or proteoliposomes into the *trans* chamber and wait for fusion to occur. Alternatively, fusion can be promoted by bringing a fire-polished glass rod dipped into ENaC-containing oocyte vesicle suspension or proteoliposomes in close proximity or even direct contact with the preformed planar bilayer membrane from the *trans* side. For this last maneuver, the bilayer membrane should be clamped at a negative voltage. Following incorporation of ENaC into the planar bilayer, channels of uniform conductance with well-defined gating transitions become apparent.
9. Voltage is manipulated and dependent currents are monitored, stored, and analyzed using a computer running *pCLAMP* software (Axon Instruments, Burlingame, CA).
10. Sensitivity to amiloride and benzamil is a hallmark of ENaC in bilayers as well as native preparations. An example of ENaC incorporated into a planar lipid bilayer is shown in **Fig. 1** (*see* **Note 12**).

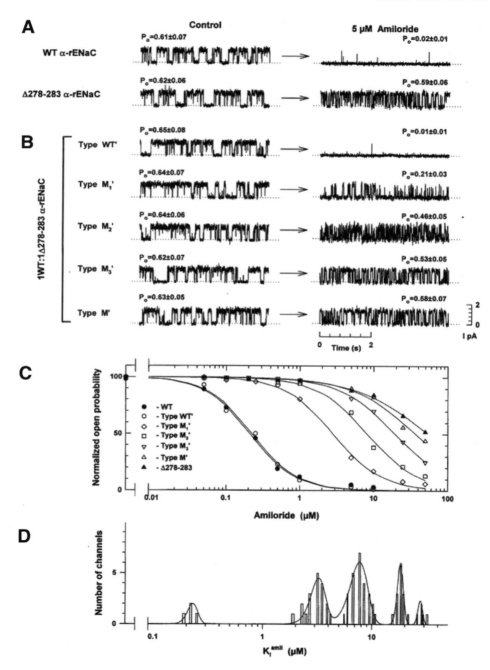

Fig. 1.

4. Notes

1. Plasmids must contain channel cDNA of interest downstream of a functional prokaryotic RNA transcription promoter, such as T7 or Sp6. In addition, having a polyA tract following the channel coding region is helpful.

2. Plasmids should be linearized to retain the promoter and polyA tail upstream and downstream, respectively, of the channel cDNA of interest. Note that distinct restriction enzymes may be required to appropriately linear different plasmids.

3. There are many different kits available for cDNA purification and in vitro transcription and translation. Use reagents which you are most familiar and comfortable. ENaC reconstitution in proteoliposomes is easily achieved with a combination of native lipids (i.e., phosphatidylethanolamine, phosphatidylserine, and phosphatidylcholine), whereas for bilayer-forming solution a mixture of synthetic lipids (i.e., diphytanoyl-phosphatidyl-ethanolamine and diphytanoyl-phosphatidylserine) is more suitable.

4. To prevent cRNA degradation, it is critical that only nuclease-free water and appropriately prepared (nuclease-free) material be used.

5. Bilayer systems can be assembled both from manufactured or homemade components *(12–14)*. Alternatively, the entire bilayer system (vibration isolation table, Faraday cage, amplifier, Bessel filter, cups and chambers, and various accessories) can be purchased from Warner Instruments.

6. Ag/AgCl electrodes embedded in agar bridges are easily prepared by "chloriding" a silver wire in bleach for 10 min, fitting chlorided wire into tubing or glass of

---

Fig. 1. Amiloride inhibition of single epithelial Na⁺ channel α-(ENaC) composed from the wild-type (WT) and amiloride-binding domain deletion mutant. The WT α-ENaC and αΔ278-283-ENaC were in vitro translated individually, mixed at a 1:1 ratio, and reconstituted into proteoliposomes in the presence of 25–50 μM DTT. Bilayers were bathed with symmetrical 100 mM NaCl solution complemented with 10 mM Tris-MOPS (pH 7.4). The holding potential was +100 mV. For illustration purposes, records were filtered at 100 Hz and are representative of at least five separate experiments with each channel. (**A**) Typical records and effect of 0.3 μM of amiloride on channels formed by the WT α-ENaC and αD278-283-ENaC alone (representative of at least five separate experiments with each channel). (**B**) Typical records of channels found in a 1:1 WT:mutant protein mixture and five typical responses of these channels to 0.3 μM amiloride: types WT' and M' were undistinguishable from the WT and mutant channels shown in **Fig. 1A**, and types M_1 through M_3 had intermediate sensitivities to amiloride. Complete amiloride dose-response curves, shown in (**C**) and determined for each population of channels shown in **B**, are representative of at least four experiments per group. Solid lines in the amiloride dose-response graph represent best fits of the experimental data points to the first-order Michaelis-Menten equation rewritten as $P_o = P_{o(max)} \cdot (1 - [\text{amiloride}]/K_i + [\text{amiloride}])$. The solid line in the statistical distribution of channel phenotypes bar graph (**D**) is a fifth-order Gaussian fit of the histogram. (Reproduced from **ref. 15** with permission from Biophysical Society.)

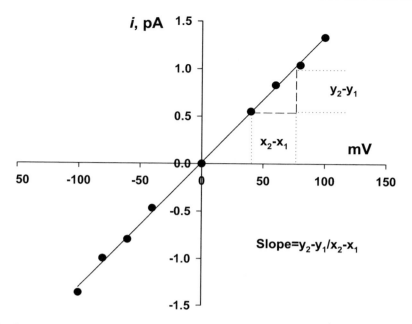

Fig. 2. Unitary current–voltage relationship of αβγ-ENaC in bilayers. Bilayers were bathed with symmetrical 100 m*M* NaCl solution complemented with 10 m*M* MOPS (pH 7.4). Conductance corresponds to the slope of the straight line.

    appropriate length and shape for a given bilayer system, backfilling the tubing or glass with freshly melted (1–3 min in microwave) 3% agar in 3 *M* KCl, and allowing the agar to solidify. The silver wire should only be about halfway down the bridge, with agar covering the wire and extending completely to the bridge end in contact with the bathing solutions.

7. Bathing solutions will vary for different channels of interest. ENaC is a Na⁺-selective ion channel; thus, our bath solutions contain Na⁺.

8. cRNA for oocyte injection is made and purified with the Ribomax Kit plus RNA cap analog and GeneClean Kits, respectively, as described for ENaC preparation for incorporation into proteoliposomes.

9. A magnetic bar and stirring device able to stir *cis* and *trans* chambers as well as a perfusion system are necessary components for a bilayer system. The bilayer bathing solutions should be made daily and filtered. The optimal temperature to perform experiments on ENaC is 25 ± 1°C. This may vary for different channels.

10. The types of lipids and lipid ratios used to form planar bilayers can be varied according to user preference and need.

11. The proper pretreatment of septum aperture with bilayer-forming solution prior to filling the *cis* and *trans* compartments of the chamber with bathing solution is helpful for bilayer formation and its stability.

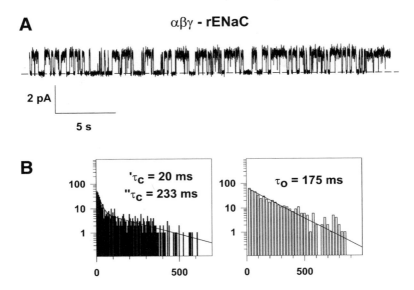

Fig. 3. Single-channel recording of αβγ-ENaC in bilayers. Bilayers were bathed with symmetrical 100 m*M* NaMOPS solution complemented with 10 m*M* Tris-HCl (pH 7.4). The holding potential was +100 mV. For illustration purposes, records representative of at least five separate experiments were filtered at 100 Hz. Representative dwell time histograms shown under the trace (**B**) were constructed following events analysis performed using *pCLAMP* software (Axon Instruments) on single-channel recordings of 3 min duration filtered at 300 Hz with an eight-pole Bessel filter prior to acquisition at 1 ms per point using *pCLAMP* software and hardware. Event detection thresholds were 50% in amplitude and 3 ms in duration. The histograms shown were constructed from 681 events. The bin widths in the closed and open time histograms were 5 and 25 ms, respectively. Closed and open time constants were determined from a double exponential $[y = a_1 \cdot \exp(-x/\tau_1) + a_2 \cdot \exp(-x/\tau_2)]$ and a single exponential $[y = a \cdot \exp(-x/\tau)]$ fit of the closed and open time histograms, respectively. (Reproduced from **ref. 16** with permission from Biophysical Society.)

12. There are a number of minimal steps that should be taken to characterize basic biophysical properties of ENaC, or any channel for that matter, after its reconstitution in bilayers. This includes, but is not limited, to the following. First, regarding *conductance*, depending on the experimental conditions, ENaC can display a closed, one or more (DTT and ultrasound-treated vs untreated) open conductive state (also referred to as a subconductance state). The conductance of open state(s) is obtained by building the current–voltage relationship of open state(s) or all-points amplitude histogram analyses. For ENaC, the current–voltage relationship usually is linear (**Fig. 2**), and its conductance corresponds to the slope of the straight line. If this relationship is not linear (not the case for ENaC under symmetrical

ionic conditions), then the conductance corresponds to the slope of the current–voltage relationship over a voltage range in which it is roughly linear. The voltage range should be specified when discussing conductance values.

Second, for *selectivity*, ENaC is a cation-selective channel. To determine the cation vs anion selectivity, a bathing solution is chosen carefully so that it includes chloride salts of a given cation (Na^+ for ENaC) that are more concentrated in one chamber than the other (e.g., 200 mM NaCl vs 50 mM NaCl). Under these conditions, the current will change sign at a voltage with a magnitude that matches the Nernst potential, depending on the cation or anion selectivity of the channel. If a given channel is permeable to both cations and anions, then the reversal potential will be less than predicted by the Nernst equation. Selectivity determination also includes the ability of the channel to discriminate between different members of the same family of ions (e.g., monovalent, divalent).

Third, *voltage dependence* refers to the time spent by a channel in the open state as a function of the applied voltage. Under symmetrical conditions, ENaC is not voltage dependent but shows some degree of sensitivity to the applied voltage under specific conditions (e.g., block by Ca^{2+}, inhibition by amiloride).

Fourth, *sensitivity* to specific pharmacological blockers helps to "positively" identify a given channel. High sensitivity to amiloride and benzamil is a characteristic of ENaC: These pharmacological "profilers" can block ENaC activity in the nanomolar range. In addition, *pCLAMP* software (Axon Instruments) allows determination of the probability of the channel being in the open state and its open- and closed-state lifetimes (often referred to as *dwell times*). Dwell times can be constructed following events list analysis by *pCLAMP* (**Fig. 3**).

Acknowledgments

ENaC cDNAs were a kind gift of Dr. B. Rossier (University of Lausanne, Lausanne, Switzerland). This work was supported by National Institutes of Health grant DK37206.

References

1. Mueller, P., Rudin, D. O., Tien, H. T., and Wescott, W. C. (1962) Reconstitution of cell membrane structure in vitro and its transformation into an excitable system. *Nature* **194,** 979–980.
2. Mueller, P., Rudin, D. O., Tien, H. T., and Wescott, W. C. (1962) Reconstitution of excitable cell membrane structure in vitro. *Circulation* **26,** 1167–1171.
3. Miller, C. (1983) Integral membrane channels: studies in model membranes. *Physiol. Rev.* **63,** 1209–1242.
4. Canessa, C. M., Horisberger, J. D., and Rossier, B. C. (1993) Epithelial sodium channel related to proteins involved in neurodegeneration. *Nature* **361,** 467–470.
5. Canessa, C. M., Schild, L., Buell, G., et al. (1994) Amiloride-sensitive epithelial Na^+ channel is made of three homologous subunits. Nature **367,** 463–467.
6. Lingueglia, E., Voilley, N., Waldmann, R., Lazdunski, M., and Barbry, P. (1993) Expression cloning of an epithelial amiloride-sensitive Na^+ channel. A new chan-

nel type with homologies to *Caenorhabditis elegans* degenerins. *FEBS Lett.* **318,** 95–99.

7. Awayda, M. S., Ismailov, I. I., Berdiev, B. K., and Benos, D. J. (1995) A cloned renal epithelial Na⁺ channel protein displays stretch activation in planar lipid bilayers. *Am. J. Physiol.* **268,** C1450–C1459.

8. Ismailov, I. I., Awayda, M. S., Berdiev, B. K., et al. (1996) Triple-barrel organization of ENaC, a cloned epithelial Na⁺ channel. *J. Biol. Chem.* **271,** 807–816.

9. Rosenberg, R. L. and East, J. E. (1992) Cell-free expression of functional Shaker potassium channels. *Nature* **360,** 166–169.

10. Perez, G., Lagrutta, A., Adelman, J. P., and Toro, L. (1994) Reconstitution of expressed K_{Ca} channels from *Xenopus* oocytes to lipid bilayers. *Biophys. J.* **66,** 1022–1027.

11. Montal, M., and Mueller, P. (1972) Formation of bimolecular membranes from lipid monolayers and a study of their electrical properties. *Proc. Natl. Acad. Sci. USA* **69,** 3561–3566.

12. Alvarez, O. (1986) How to set up a bilayer system, in *Ion Channel Reconstitution* (Miller, C., ed.), Plenum Press, New York, pp. 115–130.

13. Alvarez, O., Benos, D. J., and Latorre, R. (1985) The study of ion channels in planar lipid bilayer membranes. *J. Electrophys. Tech.* **12,** 159–177.

14. Hanke, W. and Schlue, W.-R. (1993) *Planar Lipid Bilayers*, Academic Press, San Diego, CA.

15. Berdiev, B. K., Karlson, K. H., Jovov, B., et al. (1998) Subunit stoichiometry of a core conduction element in a cloned epithelial amiloride-sensitive Na⁺ channel. *Biophys. J.* **75,** 2292–2301.

16. Berdiev, B. K., Shlyonsky, V. G., Karlson, K. H., Stanton, B. A., and Ismailov, I. I. (2000) Gating of amiloride-sensitive Na⁺ channels: subunit–subunit interactions and inhibition by the cystic fibrosis transmembrane conductance regulator. *Biophys. J.* **78,** 1881–1894.

9

A Simple In Vivo Method for Assessing Changes of Membrane-Bound Ion Channel Density in *Xenopus* Oocytes

Mouhamed S. Awayda, Weijian Shao, Ivana Vukojicic, and Abderrahmane Bengrine

Summary

Heterologous expression systems, such as *Xenopus* oocytes, are widely used to study the regulation and the structure function relationship of ion channels and transporters. In the case of ion channels, activity can be easily measured by conventional two-electrode voltage clamping. However, this method only measures the sum of the activity of all plasma membrane-bound channels. Therefore, this measurement cannot discriminate between effects on channel density and individual channel activity. To address this shortcoming, we have developed a simple assay to detect changes of membrane-bound channel density in intact oocytes. This nonradioactive assay relies on specific antibody binding in whole live cells utilizing a simple spectrophotometric measurement. This assay is linear over a wide range of channel expression levels and provides a simple cost-effective way of monitoring changes of membrane-bound channel density. Moreover, when the heterologous proteins poorly express at the plasma membrane, this method becomes advantageous to complex biochemical cell fractionation.

Key Words: Channel activity; channel density; ENaC; HRP; ion channels; immunobinding; Western blotting.

1. Introduction

Ion channels are exquisitely regulated by various second-messenger cascades to control their activity. One such example is the epithelial Na$^+$ channel (ENaC). This channel is present in various nonexcitable cells. In renal epithelia, this channel is responsible for determining distal tubule Na$^+$ reabsorption and ultimately the final urinary Na$^+$ composition (*1*). In doing so, the channel is implicated in controlling plasma volume and [Na$^+$]. Disease paradigms such as

From: *Methods in Molecular Biology, vol. 337: Ion Channels: Methods and Protocols*
Edited by: J. D. Stockand and M. S. Shapiro © Humana Press Inc., Totowa, NJ

Liddle's syndrome have validated the role of this channel in hypertension and Na⁺ homeostasis by documenting that ENaC gain of function mutations causes volume-expanded hypertension *(2)*. Thus, better understanding of the channel's regulatory processes will also improve our understanding of the physiological and pathological functions of ENaC. To accomplish this, it is necessary to determine the mechanisms of channel regulation.

Ion channels such as ENaC participate in Na⁺ movement when present at the plasma membrane. Therefore, changes of Na⁺ absorption by a cell or a coupled sheet of epithelial cells can occur via changes of channel density, channel open probability, and single-channel conductance. For the majority of nonexcitable ion channels, and certainly for ENaC, there is little or no evidence for physiologically relevant regulation at the single-channel conductance level. On the other hand, various physiological regulators and experimental maneuvers are known to affect channel density and open probability (single-channel activity) *(3–6)*. Therefore, an assessment of the experimental effects on these parameters is paramount for understanding of channel regulation and function.

Channel density and open probability have been classically measured via two separate techniques. The first method assessed changes of channel density and involved biochemical fractionation of cell membranes and assessment of protein levels via specific antibodies using Western blotting. This technique is easily adaptable to *Xenopus* oocytes *(7)*. However, in the case of ENaC and numerous other insoluble cellular proteins, the interpretation of this technique is complicated by the fact that the channel is essentially restricted to intracellular membranes with only a small amount (<5%) present at the plasma membrane (*see* **Fig. 1** and **ref. 8**). This potential pitfall is partially corrected by an additional step involving surface labeling of membrane proteins by biotin. However, this procedure adds an extra step to this protocol and is further affected by potential biotinylation of intracellular membrane-bound proteins—a major complication because the intracellular pool could be as high as 20-fold that of the plasma membrane pool.

The second technique assessed changes of single-channel activity and involved patch clamp analysis. This technique can also be used in oocytes after manual removal of the vitelline membrane *(9)*. A potential concern in these experiments is the fact that cells must undergo a shrinking-and-swelling cycle before vitelline membrane removal. Moreover, a major complication when studying ENaC is its slow gating kinetics, which preclude an accurate assessment of the open probability, especially under conditions in which it is not feasible to maintain a seal for prolonged periods. A second problem is the variable gating kinetics of ENaC, which in many instances cannot be accurately predicted by a small number of experiments (e.g., *n* = 6) *(10,11)*. Yet a third problem is a potential effect of membrane order on channel function *(12)*, which

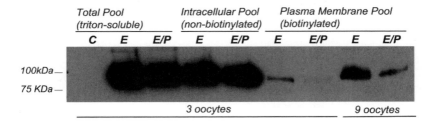

Fig. 1. Membrane and cytoplasmic ENaC pools in *Xenopus oocytes*. Proteins were probed with a anti-alpha ENaC antibody. C, E, and C/P refer to control. ENaC-expressing and PMA-treated ENaC-expressing oocytes. PMA reduced protein expression in the biotinylated pool without affecting the other pools. Note that the surface ENaC pool represents a small fraction of the intracellular membrane or total pools. Lanes were loaded with the equivalent of three or nine eggs.

may also be accompanied by effects of seal formation and membrane curvature on channel activity. Other electrophysiological techniques such as fluctuation analysis have been developed to address these shortcomings *(13)*. However, their use in oocytes is limited and is model dependent.

To circumvent these problems, we rely on a combination of two techniques: two-electrode voltage clamp and immunobinding using anti-HA or anti-FLAG antibodies in intact oocytes expressing tagged ENaC. The conventional two-electrode voltage clamp allows us to measure whole-cell currents. These currents reflect the activity of all membrane-bound ENaC (we use amiloride to block ENaC to verify this). The second measurement allows us to assess the changes of membrane-bound ENaC density. This procedure is relatively simple in that we express tagged ENaC subunits and utilize an antitag antibody already coupled to horseradish peroxidase (HRP). After binding, HRP activity can be assessed via commercial techniques (e.g., Pierce 1 Step Ultra TMB) and quantitated using spectrophotometry. The findings of this binding assay were validated in oocytes injected with different cRNA concentrations to yield an expression dose–response curve. In combination with the two-electrode voltage clamp, this technique can be used to assess effects on channel density. The protocols for this technique are described in this chapter.

2. Materials

2.1. cRNA Production

1. Linearized plasmid DNA template containing ENaC subunits (*see* **Note 1**).
2. Ribomax large-scale RNA production system T7 kit (Promega USA, Madison, WI).
3. Ribo m7G cap analog (RNA cap structure analog, New England Biolabs, Beverly, MA).

4. TE-saturated (pH 4.5; pH 8.0) phenol:chloroform:isoamyl alcohol (25:24:1).
5. Chloroform:isoamyl alcohol (24:1).
6. Nuclease-free, electrophoresis-grade agarose (Fisher Scientific, Houston, TX).
7. TAE electrophoresis buffer: 40 mM Tris-acetate and 2 mM Na$_2$EDTA at pH 8.5.
8. Denaturing gel buffer, MOPS running buffer, and formaldehyde loading dye (NorthernMax system, Ambion, Austin, TX).
9. 0.5 mg/mL ethidium bromide.
10. Isopropanol.
11. Ethanol (70%, 95%).
12. Nuclease-free water.
13. Nuclease-free 1.5-mL microfuge tubes (Fisher Scientific).

2.2. Oocyte Isolation and Culture

1. Collagenase type 1A (Sigma Chemical Co., St. Louis, MO).
2. Ca^{2+}-free OR-2 buffer: 89 mM NaCl, 2 mM KCl, 1 mM MgCl, and 5 mM, HEPES pH 7.4.
3. Rotating wheel.
4. Half-strength Leibovitz L-15 medium (Sigma) supplemented with 10 mM HEPES, 1% solution of 10,000 U penicillin/streptomycin, 50 μg/mL Amikacin free base, and 50 μg/mL neomycin sulfate at pH 7.4 with osmolarity approx 180–190 mOsm.
5. Cooled incubator or standard incubator set inside a cold room to allow 18°C.

2.3. Oocyte Expression and Recording

1. Nanoject injector (Drummond Scientific, Broomall, PA).
2. Borosilicate injecting glass, 1.5-mm diameter (Drummond).
3. Silane or SigmaCote (Sigma).
4. Fiberglass screen mesh and 5-min epoxy.
5. 10-V power supply.
6. Borosilicate recording glass, 1.7- to 1.8-mm diameter (Fisher Scientific).
7. Polyethylene tubing (PE160-190), agar, and silver wires (Sigma).
8. ND94 recording buffer: 94 mM NaCl, 2 mM KCl, 1.8 mM CaCl$_2$, 1 mM MgCl$_2$, and 5 mM HEPES, pH 7.4 and approx 190–200 mOsm osmolarity.
9. Two-electrode voltage clamp and acquisition setup (Dagan Corp., Minneapolis, MN).

2.4. Surface Biotinylation, Oocyte Fractionation, and Immunoprecipitation

1. ND94 recording buffer.
2. Biotinylation reagent: EZ-Link Sulfo-NHS-SS-Biotin (Pierce, Rockford, IL). Store desiccated at –20°C. Dissolve in ND94 immediately before each use.
3. Homogenization buffer: 20 mM Tris, 5 mM MgCl$_2$, 5 mM Na$_2$HPO$_4$, 1 mM ethylenediaminetetraacetic acid, and 80 mM sucrose, pH 7.4. Store at 4°C.
4. Protease inhibitor cocktail (Sigma).
5. TBS-Triton: 150 mM NaCl, 10 mM Tris-HCl, pH 7.5, and 1% (v/v) Triton X-100 (Sigma).

6. BCA protein assay kit (Pierce).
7. ImmunoPure immobilized streptavidin (Pierce).

2.5. Sodium Dodecyl Sulfate Polyacrylamide Gel Electrophoresis and Western Blotting

1. 4X sample buffer (Laemmli): 250 mM Tris-HCl, pH 6.8, 40% (v/v) glycerol, 8% (w/v) sodium dodecyl sulfate (SDS), 20% (v/v) 2-mercaptoethanol, and 0.1% (w/v) bromophenol blue.
2. Precast polyacrylamide gels: Ready Gel Tris-HCl (Bio-Rad, Hercules, CA).
3. Running buffer: 25 mM Tris, 192 mM glycine, and 0.1% (w/v) SDS. Do not adjust pH. Store at room temperature.
4. Prestained protein marker: Precision Plus Dual Color (Bio-Rad).
5. Transfer buffer: 25 mM Tris, 192 mM glycine, and 20% (v/v) methanol. Do not adjust pH. Store at 4°C.
6. Nitrocellulose membrane: Transfer Blot (Bio-Rad).
7. Electrophoresis and blotting paper, grade 28 (Ahlstrom, Holly Springs, PA).
8. Ponceau S protein stain: 0.1% (w/v) Ponceau S in 5% (v/v) acetic acid (Sigma).
9. TBS-Tween: 150 mM NaCl, 10 mM Tris-HCl pH 7.5, and 0.05% (v/v) Tween-20 (Bio-Rad).
10. Blocking buffer: 5% (w/v) dry milk in TBS-Tween. Dissolve milk before each use.
11. Primary antibodies: Anti-ENaC subunit-specific antibodies (Affinity Bioreagents, Golden, CO) (*see* **Note 2**).
12. Secondary antibody: HRP-conjugated goat antirabbit (Pierce).
13. ECL substrate: Supersignal West Dura (Pierce).
14. Biomax Light Film (Kodak, New Haven, CT).

2.6. HRP Binding

1. Anti-HA peroxidase high-affinity monoclonal antibody (3F10, Roche, Indianapolis, IN).
2. Anti-FLAG M2 monoclonal antibody (Sigma).
3. Antimouse immunoglobulin G conjugated to HRP (BD Biosciences, San Diego, CA).
4. HRP substrate: 1 Step Ultra TMB-ELISA (enzyme-linked immunosorbent assay) (Pierce).
5. 2 M H$_2$SO$_4$.
6. ND94 recording buffer.
7. Antibody dilution buffer: 1% (w/v) bovine serum albumin (BSA; fraction V, protease free; Sigma) in ND94 buffer.
8. 0.6-mL microfuge tubes (Fisher).
9. Transfer pipets (disposable polyethylene, Fisher).

3. Methods

Methods for the maintenance and use of *Xenopus* have been elegantly described in detail elsewhere *(14)*. In our experience, a critical part of the success of any oocyte experiment is the availability of a source of healthy, disease-resistant frogs. All of our animals are purchased from Xenopus Express (Plant City, FL). We specify South African frogs caught in the wild. Animals are primed with a small amount of human chorionic gonadotrophic hormone (50 U sc) and are used starting 2–3 wk after injection. At this low concentration of human chorionic gonadotrophin, oocyte maturity is promoted, which results in a higher percentage of oocytes in stage V/VI. Under these conditions, we rarely experience significant problems with oocyte quality.

A second, equally important, factor that affects oocyte quality is animal maintenance. All of our animals are housed in half-filled 100-gal containers. Usually, 30 animals are divided into two tanks. Animals are fed beef liver three times weekly, followed by a complete tank water change within 1–2 h after feeding. We use a backup tank that allows water to equilibrate at room temperature. After feeding, the frog-containing tank is drained, and the water from the backup tank is used to fill the tank in use. All of our frog water is tapwater filtered through a charcoal cartridge to remove chlorine and other contaminants (*see* **Note 3**).

3.1. cRNA Production

1. The linearized DNA template is cleaned using phenol followed by phenol/chloroform/isoamyl alcohol extraction. The DNA is then ethanol precipitated from the aqueous phase and resuspended in nuclease-free water. The linearized DNA template is visualized after electrophoresis on a conventional agarose gel. DNA purity and concentration are estimated from band intensity and after measuring optical density at 260 nm.

2. In vitro cRNA synthesis is then carried out using the Ribomax kit according to the manufacturer's instruction (Promega). In many instances, the amount of cap analog, or more appropriately the ratio of this analog (Ribo m7G cap analog), to guanosine triphosphate, is important to the stability of the injected cRNA and consistency of expression levels. We use a ratio as high as 5:1 and as low as 2:1. We see little effect of this ratio on the expression of rat ENaC. However, we observe large differences in the expression consistency of human ENaC.

3. At the end of the in vitro transcription reaction, the remaining DNA is digested with deoxyribonuclease. The cRNA is then extracted, ethanol precipitated, and resuspended in nuclease-free water. The integrity of the cRNA is examined by electrophoresis. In many instances, this is also verified by an in vitro translation reaction (rabbit reticulocyte cell lysate; Promega).

3.2. Oocyte Isolation, Injection, and Culture

1. Oocytes are surgically removed, followed by manual dissociation of the isolated ovarian lobes with fine forceps. This ensures more surface area contact with collagenase. To ensure prolonged viability, it is necessary to use the lowest amount

of collagenase for the shortest possible duration to defolliculate oocytes. Oocytes are first rinsed multiple times with hypertonic Ca^{2+}-free OR-2, which aids in the breakup of the tight junctions between the oocyte plasma membrane and that of the follicular cells.

2. A group of three to four ovarian lobes is incubated with approx 7 mL 1 mg/mL type 1A collagenase (*see* **Note 4**) for 2 h with one fresh solution change after the first hour. This is done at room temperature with gentle mixing in a rotating wheel. This procedure is usually sufficient to defolliculate approx 50% of the oocytes, although differences are observed with batches of collagenase and may require an adjustment for activity (*see* **Note 5**).

3. After digestion with collagenase, oocytes are placed in a 90-mm Petri dish and allowed to recover slowly. We found that gradual restoration of the $[Ca^{2+}]$ concentration is important to the long-term viability and robustness of expression. We accomplish this by dropwise addition of Ca^{2+}-containing culture medium to oocytes in the Petri dish while manually selecting stage V/VI oocytes. Newly sorted oocytes are transferred to a new, clean Petri dish with a similar $[Ca^{2+}]$. This procedure also aids in the defolliculation of the remaining cells (*see* **Note 6**). At the end of the day, stage V/VI oocytes are transferred to a new Petri dish containing the culture medium at full strength $[Ca^{2+}]$.

4. Oocytes are cultured in a modified L-15 solution. The solution is modified by dilution with water to reduce osmolarity and by the addition of 10 mM HEPES. To avoid severe bacterial contamination of this nutritive organic solution, it is necessary to use a combination of antibiotics. We find that the combination of penicillin/streptomycin, amikacin, and neomycin is sufficient to extend the usable life of these oocytes to approx 7 d in culture. This is also aided by the incubation temperature, which is controlled at approx 18°C (*see* **Note 7**).

5. Oocytes are injected with a Drummond (Nanoject, Drummond Scientific) positive-displacement, microprocessor-controlled injector. This is a relatively straightforward setup. An important requirement is that the injector glass is precoated with silane or SigmaCote to prevent cRNA loss from binding to the clean glass surfaces. A second caveat is the use of a fiberglass screen mesh (a piece of common household screen glued to the surface of a Petri dish with 5-min epoxy) to trap oocytes and prevent them from rolling around during injection. We use sharp electrodes (tips broken on the edge of a microscope slide), which yields easy impalement and further reduces the tendency of oocytes to roll around during injection. Electrode tips are kept under 20-μm diameter, which results in the least damage to the oocyte membrane while delivering consistent injections that are not blocked by intracellular yolk.

6. Oocytes are recorded from 1 to 5 d after injection. Biochemical cell fractionation and Western blotting are carried out 2–3 d after injection. The cell-binding assay is carried out on oocytes injected with cRNA coding for HA or FLAG-tagged ENaC subunits 3 d after injection.

3.3. Surface Biotinylation, Fractionation, and Streptavidin Pull-Down

1. Oocytes are injected with 2–5 ng cRNA for each ENaC subunit. Oocytes are used 2–3 d after injection.

2. Oocytes are placed in 60-mm Petri dishes and are washed three times in ND94 buffer on ice. Groups of 50 oocytes are then transferred to 1.5-mL microcentrifuge tubes containing a total volume of 1.35 mL ND94.

3. The biotinylation reagent (sulfo-NHS-SS-biotin) is freshly dissolved in ND94 buffer at a concentration of 10 mM (6 mg/mL), and 150 µL of this stock solution is added to the oocyte-containing tubes to a final concentration of 1 mM. This crosslinking reagent is cleavable by reducing agents and allows examination of the electrophoretic mobility of the protein of interest (ENaC) in the absence of additional contribution from the coupled biotin molecules. Tubes are incubated with this reagent at 4°C with gentle rotation (~6 rpm) for 45 min.

4. At the end of the incubation period, oocytes are washed four times with cold ND94 and are incubated in homogenization buffer, with 1% protease inhibitor cocktail, at a volume of 10 µL/egg. Oocytes are homogenized on ice by passing six times through a 23-gage needle and another six times through a 27-gage needle.

5. The homogenate is centrifuged for 5 min at 4°C and 200g to pellet the yolk.

6. The resulting supernatant (yolk-free homogenate) is then centrifuged for 20 min at 4°C at 14,000g to pellet the membrane fraction.

7. The 14,000g-fraction supernatant (water-soluble fraction) is not used any further (*see* **Note 8**). The pellet ("membranes") is resuspended in TBS-Triton (10 µL/egg), incubated on ice for 30 min, and the suspension is then centrifuged for 20 min at 4 EC, at 14000g.

8. The new supernatant (Triton-soluble membrane fraction) was previously found in Western blotting to contain the entire ENaC pool and is therefore used in further experiments. On the other hand, no ENaC was detected in either the water-soluble or Triton-insoluble fractions.

9. Protein concentrations of the Triton-soluble fractions are determined using the BCA protein assay kit. Solubilizing membranes in volumes of 10 µL/egg typically yielded protein concentrations of approx 0.4 mg/mL.

10. To isolate biotinylated membrane protein, 150 µL of the streptavidin agarose suspension is added. The mixture is then incubated for 1 h at 4°C with rotation. The agarose beads are then collected by brief centrifugation at 10,000g.

11. The supernatant (assumed to contain nonbiotinylated proteins) is saved for Western blotting, and the agarose beads (with bound biotinylated proteins) are washed three times with 1 mL TBS-Triton.

12. After the last wash, the buffer covering the agarose beads is removed as completely as possible using a 27-gage needle attached to a 1-mL syringe. SDS-PAGE (polyacrylamide gel electrophoresis) sample buffer is then added to the beads, and the samples are heated at 90°C for 5 min to release the biotinylated surface proteins. The aqueous phase of the bead solution is then collected and used for Western blotting.

3.4. SDS-PAGE and Western Blotting

1. Sample buffer is added directly to agarose beads as described in **Subheading 3.3.**, **step 12**. For samples in solution, one-third volume of 4X sample buffer is added. Protein samples are then prepared for SDS-PAGE by heating for 5 min at 90°C.

2. SDS-PAGE is carried out using the Mini Protean II System from Bio-Rad and utilizing precast polyacrylamide gels (4% stacking gel, 7.5% separating gel).

3. Samples and the prestained protein markers (8 μL) are loaded and electrophoresed at 150 V for approx 1 h (until the bromophenol blue dye front reaches the bottom of the gel).

4. At the end of the electrophoresis period, the gels are removed, and the upper right corner of the gel and transfer membrane are cut off to allow us to follow the orientation of the gel. Protein transfer is carried out in 20% methanol transfer buffer using the Mini Trans-Blot cell from Bio-Rad according to the manufacturer's instructions.

5. Transfer is carried out for 1.5 h at 100 V. After disassembling the transfer sandwich, the prestained protein markers are now absent from the acrylamide gel but easily visible on the transfer membrane. This provides a positive indication of successful transfer.

6. After the transfer, membranes are incubated in a Ponceau S solution for 5 min to visualize the protein bands and verify the quality and uniformity of protein separation and transfer. Membranes are then rinsed with water several times to remove the Ponceau stain (any traces of red color remaining after repeated washes will be removed in the next step).

7. Membranes are incubated in blocking buffer (30 mL per blot) for 1 h at room temperature or overnight at 4°C with shaking.

8. The blocking buffer is then replaced by 5 mL of the primary antibody solution diluted 1:2000 in blocking buffer (final concentration of 0.5 μg/mL). The blot is incubated in this solution for 1 h at room temperature with shaking.

9. The primary antibody is then removed, and the membranes are washed five times for 5 min each with TBS-Tween (30 mL per wash). This is followed by addition of 5 mL of the secondary antibody solution (diluted 1:500 in TBS-Tween; final concentration 20 ng/mL) for 1 h at room temperature with shaking.

10. The secondary antibody solution is removed, and the membranes are washed five times for 5 min each with TBS-Tween (30 mL per wash). During the washes, 1 mL each of components A and B of the ECL substrate are warmed to room temperature and mixed (total 2 mL substrate per blot).

11. After the last wash, the membranes are rinsed with TBS (no Tween) and drained. The remaining buffer is wicked away with a Kimwipe. The mixed ECL substrate is then poured over the membrane.

12. After 5 min of incubation at room temperature, the membranes are drained, and the remaining substrate is wicked away. The membranes are then wrapped in clear plastic film and exposed to film in an autoradiography cassette. Film exposure times are adjusted for optimal signal visualization. Film is developed using an automated developer. Data are scanned using a flatbed digital scanner. An ex-

ample of the biotinylation and Western protocol is shown in **Fig. 1**. Included are the effects of PMA, which are known to involve ENaC endocytosis from a membrane pool *(15)*.

3.5. Binding Assay

1. Oocytes are injected with cRNA for either FLAG- or HA-tagged ENaC subunits. Those expressing HA-tagged ENaC utilize an HRP-conjugated primary antibody and therefore do not require the use of a secondary. Those expressing FLAG-tagged ENaC require the use of an HRP-conjugated secondary antibody. We inject 2–5 ng of each subunit cRNA. Binding is carried out 3 d after injection. This results in the highest electrophysiological and biochemical signals.

2. The day before the binding experiments, 0.6-mL tubes are coated overnight with the antibody-diluting buffer (containing BSA) at 4°C. This procedure minimizes the nonspecific binding of the antibodies to the tube material.

3. On the day of the experiment, oocytes are examined under a microscope. Atretic oocytes, those with marbleized membranes, or those with visible blebbing are discarded (usually <5% of all injected oocytes). Oocytes are washed three times with ND94 solution and are sorted into groups of 25. All control and experimental groups are carried out in triplicates or quadruplicates. Data are averaged for each group and treated as a single experiment.

4. Utilizing the precoated 0.6-mL tubes, the appropriate antibodies (anti-HA or anti-FLAG M2) are diluted 1:70 into 0.5 mL of the antibody dilution buffer (final concentration 0.357 µg/mL and 62.8 µg/mL for the HA and FLAG antibodies, respectively).

5. Using a transfer pipet, each group of 25 oocytes is gently transferred into the 0.6-mL tube containing the appropriate prediluted antibody. Tubes are then placed on a rotating wheel set at low speed (~6 rpm). Antibody binding is carried out for 90 min at room temperature (*see* **Note 8**).

6. The binding reaction is terminated by placing the tubes on ice. A transfer pipet is then used to remove the majority of the antibody-binding solution (leaving just enough to prevent oocyte lysis), followed by extensive washing.

7. Oocytes are washed three times in the 0.6-mL tubes with ND94 solution. Oocytes are then transferred to 15-mL tubes and washed three times with 10 mL ND94. All washes are carried out on ice. The HRP activity of the last wash solution is measured to confirm the effectiveness of the washes. Usually, this results in optical density values of less than 0.05.

8. In experiments utilizing the HA-tagged HRP-conjugated antibody, washed oocytes are transferred to 1.5-mL tubes containing 100 µL HRP substrate (1 Step Ultra TMB). The reaction is allowed to proceed at room temperature for 15 min. At this point, the signal can be observed as a visible blue color. This signal can then be measured at an optical density of 375 or 656 nm; however, at this point the signal is unstable for prolonged periods and is also susceptible to degradation (*see* **Note 9**).

9. A 75-µL aliquot of the blue solution is transferred to a new tube containing 75 µL 2 M H_2SO_4. This results in a color change to yellow that can be measured at a

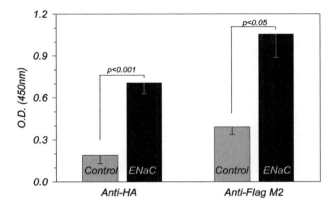

Fig. 2. Antibody binding in control and ENaC-expressing oocytes. Experiments were carried out as described in the binding assay in the text. The resolution of both the anti-HA and anti-FLAG antibodies are compared. While both antibodies generated a much larger signal in ENaC-expressing versus control oocytes, the anti-HA antibody exhibited a lower background signal and a higher control to background ratios ($n = 5$).

45-nm OD. The values obtained at this wavelength result in improved signal-to-background ratios. Moreover, the solution is now stable for extended periods (>24 h), which allows the convenience of measuring absorbance at a later date.

10. Usually, this entire procedure can be accomplished with little or no loss of oocyte membrane integrity. However, some signal interference is observed in groups with broken oocytes (cloudy solution). This can be avoided by centrifuging the tubes at high speed prior to measuring optical density.

11. Optical density is measured in a 100-μL cuvet. A mixture of equal volumes of HRP substrate and 2 M sulfuric acid is used as a blank. Solutions with OD > 1 are remeasured after a 1:2 or 1:4 dilution to avoid problems with signal saturation and nonlinearity.

12. Oocytes utilizing the anti-FLAG antibody follow the same procedure as that outlined in **steps 1–7** except that oocytes are washed three times in the 0.6-mL tubes to remove the primary antibody. Oocytes are then transferred to another 0.6-mL tube (blocked with BSA overnight) containing 0.5 mL the secondary antibody (antimouse immunoglobulin G diluted 1:500 in antibody dilution buffer; final concentration 2 μg/mL).

13. At the end of the 60-min incubation period, oocytes are treated as those incubated with the anti-HA antibody starting in **step 7**. As this protocol utilizes an additional binding step, it is not surprising that it results in higher background signals. An example of the result observed with this binding assay is shown in **Fig. 2**. This figure also compares the signal-to-background values in control and ENaC-expressing oocytes utilizing the single-step HA and the two-step FLAG protocols.

14. To determine the dynamic range of this assay and to verify the capacity to detect changes of channel density, oocytes can be injected with various amounts of

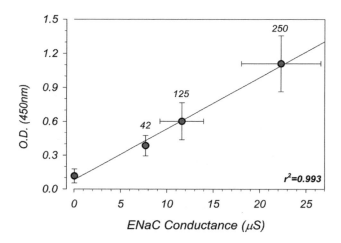

Fig. 3. Anti-HA binding at different ENaC expression levels. Oocytes were injected with 42, 125, or 250 ng of ENaC cRNA. An electrophysiological index of expression was assessed from the amiloride sensitive conductance, while a biochemical index of membrane density was assessed from the binding assay using the HA-tagged protocol. A linear relationship is observed with an r2 of 0.993, indicating the ability of this assay to detect changes of membrane bound ENaC density over a wide range of expression.

ENaC cRNA. This results in different expression levels, which can then be used to correlate with antibody binding. An example of this is shown in **Fig. 3**. This figure demonstrates that this binding assay is linear over a wide range of ENaC expression.

3.6. Two-Electrode Voltage Clamp

1. Two-electrode voltage clamp is carried out by impalement with two intracellular microelectrodes. These electrodes are conventional 3 M KCl borosilicate electrodes. Electrodes are pulled in three stages to ensure small tip diameters with relatively low access resistance. Electrode resistance is usually in the range of 1 to 2 MW.

2. Bath reference electrodes are also utilized and are constructed from Ag/AgCl electrodes connected to the chamber via 2.5% agar bridges in 3 M NaCl. The formation of the silver chloride wire is accomplished by imposing a direct current holding voltage of 7 V for approx 30 s between two pure silver wires in a solution of 0.1 N HCl. This results in the formation of AgCl in one of the wires and hydrogen gas in the second. These Ag/AgCl electrodes form the connection between the bath clamp unit of the amplifier and the agar bridges. The agar bridges are constructed by trapping a hot solution of liquified agar containing 3 M NaCl in a short piece (~2 cm) of polyethylene tubing (PE 160–190). These bridges connect the chamber to the Ag/AgCl wires. This setup avoids problems with electrode polarization and allows chamber perfusion with solutions of varying [Cl$^-$].

3. Oocytes expressing ENaC exhibit an open circuit voltage in the range of −10 to +20 mV. This is a reflection of the new Na^+ equilibrium potential and depends partly on the degree of ENaC expression. To avoid drifts and changes of intracellular Na^+ concentration, the membrane potential is clamped to 0 mV (*see* **Note 10**). At this voltage, the whole-cell current is usually close to 0 mV and is an inaccurate measure of ENaC activity. Therefore, the use of current as an estimate of ENaC expression should be avoided.

4. Whole-cell channel activity is determined by measuring conductance *(15)*. Amiloride is added at 10 μ*M* to block more than 98% of ENaC activity and to allow subtraction of the contribution of oocyte endogenous background conductances. The amiloride-sensitive conductance is measured in the same group of oocytes as those used for binding experiments. These measurements are described in detail elsewhere *(15)* and yield an estimate of channel activity unaffected by the holding voltage or the $[Na^+]_i$.

4. Notes

1. DNA template is linearized by digestion with an appropriate endonuclease. It is recommended to use enzymes that produce a 5′ overhang. If an enzyme producing a 3′ overhang must be used, then the linearized DNA should be blunt ended before its use in a transcription reaction to avoid the formation of extraneous transcripts.

2. Equal volume of glycerol was added to the antibody solutions, which allows for storage at −20°C in the original vial and avoids freeze-thaw cycles. We found this more beneficial for preserving the antibodies' reactivity than the method recommended by the manufacturer (diluting the antibody with BSA-containing buffer and freezing single-use aliquots).

3. The quality of water is critical to the quality of isolated oocytes. It is important to avoid distilled water as it has been our experience that frogs are visibly stressed in such water, and many eventually do not survive for prolonged periods. We use tapwater that is passed through a large (100-cm) charcoal cartridge. We routinely test the chlorine content (biweekly). It is important to employ such a regiment because often oocyte quality can be traced to high levels of chlorine in the water. It should be pointed out that in this case animals continue to look healthy, and thus their appearance cannot be used to predict chlorine water problems.

4. We have tested several types and purity levels of collagenase. We found that type 1A, which contains clostripain, neutral proteases, and tryptic activities, is the best suited for follicular cell removal and long-term oocyte survival in culture. For example, the lot we are currently using contains 409 U/mg collagenase, 44 U/mg neutral protease (caseinase), and 0.54 U/mg clostripain.

5. A reliable index of whether the collagenase concentration or duration of incubation needs adjustment can be observed after the first hour of treatment. A small sample of oocytes is examined after the first 60 min of digestion. If the digestion concentration is appropriate, then we observe that the oocytes are completely dissociated from their ovarian lobes during this period. During the second 60 min of digestion we observe that approx 50% of the oocytes are either completely or

partially defolliculated. Removal of the remaining follicular cells is then accomplished nonenzymatically (*see* **Note 6**).

6. The membranes of follicular cells will adhere to the surface of a clean Petri dish. This is observed within 10–15 min after placing partially defolliculated oocytes in a fresh Petri dish. Oocytes can then be removed from their remaining attachment to the follicular cells by manually swirling the Petri dish. This simple procedure allows us to use a submaximal concentration of collagenase. If necessary, we repeat this procedure a second time.

7. Isolated oocytes are visibly healthy for 1–2 d after defolliculation irrespective of the incubating solution (including the ND94 recording solution). However, in the absence of any solution modification, the animal and vegetal poles of oocytes begin to mix (marbleization). This is accelerated at room temperature (>20°C). The membranes of these oocytes become extremely fragile and in many instances develop blebbing that can be observed under high magnification (×50). This is also observed electrophysiologically, for which the conductance of control or amiloride-treated ENaC-expressing oocytes becomes unusually high. We find that the addition of the nutritive media along with the incubation at 18°C extends the usable life of these oocytes to approx 1 wk. Moreover, this allows gradual expression changes and avoids a large spike in expression levels at 1–2 d after injection.

8. Avoid trapping air bubbles when adding oocytes to the antibody-containing tube. The angle of the rotating wheel may also require adjustment to avoid trapping air bubbles. Oocytes should be freely moving within the microfuge tube, or the trapped air can cause poor contact with the antibody solution and can increase the risk of oocyte membrane breakdown.

9. It is important not to allow the color reaction to proceed to completion. We found that 15 min was an appropriate time. If the reaction is allowed sufficient time to proceed to completion, then it may not be possible to differentiate between different expression levels. Moreover, the instability of the initial reaction (the one prior to the addition of the H_2SO_4) also factors into this problem. This is observed as an additional color change from the blue to the yellow direction in the absence of sulfuric acid. This is usually observed if the reaction is allowed to proceed for more than 45 min.

10. A more appropriate protocol would have been to clamp to the open-circuit voltage or to hold at positive voltages such as +20 mV to prevent Na^+ loading and inhibition of ENaC. However, this is not possible in oocytes because prolonged holding at positive voltages causes time-dependent stimulation of an endogenous cation channel (*16*). This is never observed at 0 mV. Thus, this voltage is the best possible compromise and allows us to carry out prolonged experiments without channel rundown. Indeed in many long experiments (>60 min) we even observe a small, time-dependent increase of activity.

Acknowledgment

This work was supported by National Institutes of Health grant DK55626.

References

1. Garty, H. and Palmer, L. G. (1997) Epithelial sodium channels: function, structure, and regulation. *Physiol. Rev.* **77**, 359–396.
2. Rossier, B. C., Pradervand, S., Schild, L., and Hummler, E. (2002) Epithelial sodium channel and the control of sodium balance: interaction between genetic and environmental factors. *Annu. Rev. Physiol.* **64**, 877–897.
3. Rotin, D., Kanelis, V., and Schild, L. (2001) Trafficking and cell surface stability of ENaC. *Am. J. Physiol. Renal Physiol.* **281**, F391–F399.
4. Rossier, B. C. (2002) Hormonal regulation of the epithelial sodium channel ENaC: N or P(o)? *J. Gen. Physiol.* **120**, 67–70.
5. Rossier, B. C. (2003) The epithelial sodium channel (ENaC): new insights into ENaC gating. *Pflugers Arch.* **446**, 314–316.
6. Stockand, J. D. (2202) New ideas about aldosterone signaling in epithelia. *Am. J. Physiol. Renal Physiol.* **282**, F559–F576.
7. Kamsteeg, E. J. and Deen, P. M. (2001) Detection of aquaporin-2 in the plasma membranes of oocytes: a novel isolation method with improved yield and purity. *Biochem. Biophys. Res. Commun.* **282**, 683–690.
8. Firsov, D., Schild, L., Gautschi, I., Merillat, A. M., Schneeberger, E., and Rossier, B. C. (1996) Cell surface expression of the epithelial Na channel and a mutant causing Liddle syndrome: a quantitative approach. *Proc. Natl. Acad. Sci.* **93**, 15,370–15,375.
9. Methfessel, C., Witzemann, V., Takahashi, T., Mishina, M., Numa, S., and Sakmann, B. (1986) Patch clamp measurements on *Xenopus laevis* oocytes: currents through endogenous channels and implanted acetylcholine receptor and sodium channels. *Pflugers Arch.* **407**, 577–588.
10. Palmer, L. G. and Frindt, G. (1996) Gating of Na channels in the rat cortical collecting tubule: effects of voltage and membrane stretch. *J. Gen. Physiol.* **107**, 35–45.
11. Awayda, M. S. and Subramanyam, M. (1998) Regulation of the epithelial Na$^+$ channel by membrane tension. *J. Gen. Physiol.* **112**, 97–111.
12. Awayda, M. S., Shao, W., Guo, F., Zeidel, M., and Hill, W. G. (2004) ENaC–membrane interactions: regulation of channel activity by membrane order. *J. Gen. Physiol.* **123**, 709–727.
13. Segal, A., Awayda, M. S., Eggermont, J., Van Driessche, W., and Weber, W. M. (2002) Influence of voltage and extracellular Na(+) on amiloride block and transport kinetics of rat epithelial Na(+) channel expressed in *Xenopus* oocytes. *Pflugers Arch.* **443**, 882–891.
14. Kay, B. K. and Peng, H. B. (eds.). (1991) *Methods in Cell Biology, Vol. 36:* Xenopus laevis: *Practical Uses in Cell and Molecular Biology.* Academic Press, New York.
15. Awayda, M. S. (2000) Specific and nonspecific effects of protein kinase C on the epithelial Na(+) channel. *J. Gen. Physiol.* **115**, 559–570.
16. Weber, W. M. (1999) Endogenous ion channels in oocytes of *Xenopus laevis*: recent developments. *J. Membr. Biol.* **170**, 1–12.

10

Preparation of Cortical Brain Slices for Electrophysiological Recording

Costa M. Colbert

Summary

Acute brain slices allow electrophysiological and imaging techniques to be applied in vitro to the study of neuronal ion channels, synaptic plasticity, and whole-cell function in juvenile and adult tissue. Ion channel recordings from small dendritic branches and axons in brain slices have demonstrated considerable functional differences in ion channel function within subregions of single cells. These findings have greatly increased our understanding of neuronal computation. This chapter presents methods for obtaining high-quality brain slices developed to aid visualization of small neuronal structures using differential interference contrast microscopy.

Key Words: Brain slice; fluorescence imaging; hippocampus; ion channels.

1. Introduction

Acute brain slices allow neural tissue that has developed normally in vivo to be used as an in vitro preparation. The primary reason for the relatively thin slices (250–500 μm) is to allow adequate oxygen to reach the cells without an intact cardiovascular system. However, slices also allow microscopic visualization of neurons, relatively rapid drug application, and electrical stimulation that would not be possible in vivo.

Brain slices have been used since the mid-1950s for studies of hippocampus and other cortical and subcortical regions (*see* **ref. 1**). By today's standards, many of the earliest slices were of dubious quality in terms of overall health. However, by the early 1980s, high-quality extracellular field studies were commonplace, primarily in studies of synaptic plasticity. Intracellular recordings using sharp electrodes were less common but were well within the capabilities

From: *Methods in Molecular Biology, vol. 337: Ion Channels: Methods and Protocols*
Edited by: J. D. Stockand and M. S. Shapiro © Humana Press Inc., Totowa, NJ

of many laboratories studying modulation of neuronal excitability, dendritic integration, and to a lesser degree, ion channel kinetics. The advent of the blind patch recording technique *(2)*, by virtue of its lower noise level, allowed the use of whole-cell recordings in slices to investigate synaptic transmission in the central nervous system.

In the early 1990s, the application of differential interference contrast microscopy to brain slices allowed direct visualization of live, unstained neurons within the slices *(3)*. This technique made patch recording somewhat easier but, more important, allowed the entire array of patch clamp recording techniques to be applied in slices. More importantly, visualization of dendrites and axons allowed site-specific recordings to be made *(4)*. This capability allowed differences in ion channel kinetics and distribution to be identified within single neurons (e.g., **ref. 5**). It also brought about a need for exceptionally high-quality slices in which these structures could be readily identified.

The approaches described in this chapter were developed to allow visualization of small structures in slices from adult animals. They involve choice of solutions, perfusion of the animal with cold saline to cool the brain rapidly before the supply of oxygen is interrupted, choice of slice orientation, slicing itself, and maintenance of the slices before and during recording.

2. Materials

2.1. Solutions

1. Perfusion and cutting solution: 110 mM sucrose, 60 mM NaCl, 1.25 mM NaH$_2$PO$_4$, 3 mM KCl, 28 mM NaHCO$_3$, 7 mM MgCl$_2$, 5 mM dextrose, 1.3 mM ascorbate, 2.4 mM pyruvate, and 0.5 mM CaCl$_2$.
2. Holding solution: 125 mM NaCl, 2.5 mM KCl, 1.25 mM NaH$_2$PO$_4$, 25 mM NaHCO$_3$, 1 mM MgCl$_2$, 25 mM dextrose, 2 mM CaCl$_2$, 1.3 mM ascorbate, and 2.4 mM pyruvate (*see* **Note 1**).

2.2. Dissection, Slicing, and Perfusion Equipment

1. Perfusion cannula: a gravity-fed perfusion system is easily constructed from an intravenous catheter kit and a 30-mL syringe. The syringe is connected to the tubing above the drip chamber. The syringe acts as the reservoir for the perfusion solution, and the intravenous kit allows the flow to be regulated. The syringe is hung about 2–3 ft above the level of the animal. A syringe needle (16–20 gauge) or a glass pipet is attached to the end of the intravenous tubing to pierce the heart.
2. Toothed forceps, strong sturdy scissors, small scissors sturdy enough to cut the skull, scalpel handles and blades, and small, flat weighing spatula; wide-mouth pipet, either glass or plastic; two syringes with 25-gauge needles, one bent to about 60°, to trim and move slices within the specimen tray; Superglue, either 3M tissue glue or Locktite 404 works well, but note that each requires Ca^{2+} to

harden, and thus it will sit as a liquid until the brain (or your finger) comes in contact.

3. Vibrating blade tissue slicer: a number of offerings are available on the market, including the Vibratome, sold by a number of distributors, and those from Leica, Ted Pella, Cambridge, and others. The tissue slicer should have independent control of vibration amplitude and cutting speed.

4. Specimen tray: we use a custom-made specimen tray in a Vibratome to hold the brain during cutting. The tray's central block is made of stainless steel. This block allows the tray to be securely fastened in the Vibratome's tray vise, provides a smooth surface that can easily be cleaned with a razor blade, and helps keep the brain cold. We keep the tray in the freezer before cutting. The rest of the tray, made from Lexan, is just deep enough to allow the brain to be fully submerged. The dimensions of the tray are somewhat larger than the standard specimen tray sold as an option for the Vibratome to allow for a larger volume of cutting solution.

5. Blades: there are a number of blades that can be used with a vibrating tissue slicer. Conventional razor blades are suitable for very young tissue only. Carbon "blue" blades from Japan ("Feather" blades, distributed by Ted Pella and others) produce better slices but are still limited to young tissue. For older animals, say past 4 wk for rat slices, the use of sapphire or glass blades (Delaware Diamond Knives, Ted Pella) provides more consistent results. These blades are expensive and can be sharpened only once or twice. However, unless they are mistreated, they can be used daily for 12–18 mo. They work fine even when there are microscopic nicks in the blade. We inspect the blade periodically, mark bad sections with a Sharpie pen, and arrange the blade and brain so that these sections are avoided. Some labs use glass microtome knives. These are made daily using a knife breaker. The major drawback is the amount of effort needed to make good blades, which can be used a few times at most.

6. An osmometer (vapor pressure type) is not absolutely necessary but is useful for checking that solutions have been made correctly. Extracellular solutions should be about 310–320 mOsm (*see* **Note 2**).

7. Slices can be maintained for many hours at room temperature using a submerged chamber. The key is to have the solution continuously and well bubbled with oxygen/carbon dioxide and for there to be substantial flow. We use a chamber made from a 150-mL beaker, a soft plastic disposable pipet, and a 25-mm plastic Petri dish (**Figs. 1** and **2**). A Lexan disk just fits the diameter of the beaker. In this disk are two cutouts, one that just fits the Petri dish and one that just fits the broad part of the pipet. The pipet is cut top and bottom to produce a small chimney. The Petri dish is cut out so that all that is left is a ring. A single layer of a cotton 4 × 4 swab is sandwiched between this ring and the Lexan disk to provide a sparse, but tight, net to support the slices. A bubbler is placed in the pipet chimney. As the bubbles rise, they impart a slow flow to the solution, which then is gently forced across the slices.

Fig. 1. Side view of slice-holding chamber. Chamber consists of a platform for holding slices and a chimney made from a plastic disposable pipet. Solution completely covers the platform and the top of the chimney. A bubbler in the chimney oxygenates the solution. Bubbles in the chimney produce a gentle flow of solution past the slices.

3. Methods

3.1. Solution Preparation

1. Solutions should be made in 1- or 2-L volumes, depending on use. Solutions should be kept refrigerated. Do not use solutions more than a few days old (*see* **Note 3**).
2. Add the $CaCl_2$ last, after bubbling the solution with carbogen (95% O_2/5% CO_2) for a couple of minutes (*see* **Note 4**).

3.2. Perfusion

1. Cool perfusion solution by placing in the freezer until some ice forms. Mix well. A "slushy" blender can be used to make a slush, but there should be no ice by the time the perfusion begins (*see* **Note 2**).

Fig. 2. Top view of slice-holding chamber. Note the Lexan insert with holes for the chimney and the slice platform. A drilled-out plastic Petri dish mates with the insert to lock down a single layer of cotton 4 × 4 to provide a broad net to hold the slices without impeding solution flow through the slices. The extra holes in the insert are decorative.

2. Fill reservoir with about 30 mL cold perfusion solution. Bubble with carbogen (95% O_2/5% CO_2). Allow the solution to flow to the end of the cannula, then clamp. Be sure there are no bubbles in the tubing.

3. Anesthetize animal deeply with approved anesthetic.

4. Once deep anesthesia is reached, open the abdominal cavity using scissors, exposing the diaphragm. A pair of *toothed* forceps should be used to hold the tissue while cutting with the scissors.

5. Cut the diaphragm to expose the heart. Cut the ribcage laterally to allow easy access to the thoracic cavity. The xyphoid process makes a convenient, sturdy handle to secure the ribcage.

6. Place the tip of the perfusion cannula into the left ventricle. If a glass cannula is used, then knick the ventricle with a scalpel first. The advantage of the glass cannula is that it can be moved into the aorta easily without popping through the heart wall.

7. Lift the heart up and knick the right atrium, preferably at the auricle.
8. Open the clamp on the perfusion line; adjust the cannula until good flow is achieved, then secure the cannula. There should be a continuous stream of solution flowing in the drip chamber of the intravenous set. If there are only a few drops per second, then the cannula should be readjusted. Constantly monitor the flow.
9. The paws and eyes should clear quickly. Allow the bulk of the solution to flow through.
10. Remove the head in preparation for removing the brain. Submerge the head in an ice-cold beaker of cutting solution (150 mL, half filled).

3.3. Removing the Brain

1. During removal of the brain, try to keep the head cold. We place a piece of filter paper in a 100-mm Petri dish, place the head on the paper, and occasionally add more ice-cold solution on top of the head.
2. Using a scalpel, make a midline cut to the bone from just above the eyes to the back of the head. Pull away any loose skin. If the perfusion was successful, then everything should be pale, and there should be very little blood in the skin and muscle.
3. With a small pair of scissors, cut from the foramen magnum along the midline to just past bregma (the anterior suture line crossing). Keep the tips pointed up, away from the brain. Make a cut laterally along the posterior suture line (*see* **Note 5**).
4. Using Rangeurs, pull the skull laterally on each side to expose the brain. Avoid breaking off small pieces of bone, which can be very sharp.
5. Carefully lift the skull away from the brain, taking care not to poke or to apply pressure to the brain.
6. Reach down laterally to the cerebellum and crush the temporal bones on each side to allow easier brain removal.
7. With a scalpel, make a cut laterally on each side at about the level of bregma.
8. Lift the brain with a small spatula. Cut the optic nerves and the trigeminal nerves as you lift the brain out of the skull. It is important to be gentle and not to stretch or compress the brain as you lift it out of the skull. Place the brain in cold perfusion solution.

3.4. Blocking the Brain

1. Once the brain has been removed, it should be placed on a piece of filter paper in a Petri dish and covered with a small amount of cold solution. At this time, retrieve the specimen tray from the freezer.
2. Using a sharp clean scalpel, cut along the midline into two hemispheres. On each side of the brain, cut a small amount off the dorsal aspect of the brain so that the brain half can sit stably on its dorsum.
3. Dry off the brain by setting it gently onto dry absorbent paper.
4. Glue the brain to the stage with the posterior end facing the blade on the tissue slicer (**Fig. 3**). If desired, repeat with the other side of the brain. This is faster but

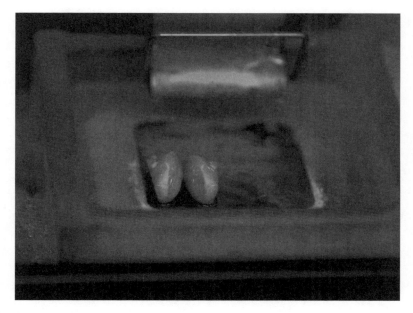

Fig. 3. Brain blocked and glued to specimen tray of tissue slicer. For hippocampal slices, brain is oriented so that the entorhinal cortex faces the oncoming blade. The brain has been glued to the side to avoid a rough spot on the sapphire blade.

not necessary. The sides can be cut sequentially without any problems with viability (*see* **Note 6**).

5. Trim any pia that remains attached to the brain. This is evident if the slices curl as they are cut. A small set of forceps is useful for this task. Good lighting is essential. A dissecting scope can also help if pia is a problem, which it tends to be with older animals.

3.5. Slicing

1. Amplitude should generally be high and speed of cutting low. Some trial and error is necessary because of differences between the instruments. Typically, 300- to 400-μm slices work best.
2. Because the desired slices are often well below the surface, it helps to keep track of the depth from the surface of the brain. For example, we often make 25 turns of the depth dial before starting to look carefully at the slices.
3. Just before the depth of the slices of interest, make a vertical cut through the tissue with a razor blade. Be careful not to move or stretch the brain, or it may come unglued from the base. This cut allows the slices to release from the brain before the blade gets to the end of its travel. If this is not done, then the slices have a tendency to flip at the end and not release cleanly (**Fig. 4**).

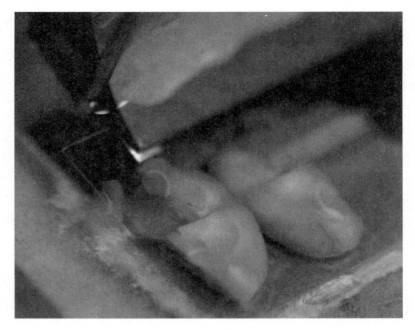

Fig. 4. Horizontal hippocampal slices released from block of brain. A vertical cut has been made in the block of brain to allow the slice to release cleanly. The forward motion of the blade is stopped once the slices release to preserve the stopping function of the anterior portion of the brain. Excess tissue can be trimmed from the slice while it sits atop the blade.

4. Once the slice releases, stop the forward travel of the blade.
5. Trim any excess tissue from the slice. Be careful not to drag the slice around the tray while cutting.
6. Use a wide-mouth pipet to transfer the slice. Allow the slice to float down into the holding chamber but do not allow all the solution to go into the holding chamber; otherwise, after 10 or so transfers you will have added a significant extra volume (of cutting solution) to the holding chamber (*see* **Note 7**).
7. Slices should be maintained in the holding chamber for 10–20 min at about 31°C. Afterward, the holding chamber should be held at room temperature (*see* **Note 8**).

4. Notes

1. Solutions based on choline chloride have been used instead of sucrose. In both cases, the idea is to limit excitability by lowering Na^+ concentration. However, choline is a partial agonist at cholinergic receptors and may have a number of effects. Neurons and slices maintained in choline chloride have a resting potential around –30 mV. Although this is not the case if the slices are only exposed to

choline during perfusion and slicing, it remains worrisome, and we no longer advocate the use of choline.

2. Osmolarity of extracellular solutions is typically 290–300 mOsm. Frozen solutions with ice crystals will often have much higher osmolarities because the water freezes without including the salt. High osmolarity (>350–400) easily occurs and will kill brain tissue.

3. Mix solutions in about two-thirds of the final volumes. Bring volume up to the final volume after salts are dissolved. The salts (and sucrose in particular) have a high enough concentration to displace a lot of water. If you start with the final volume of water, then you will end up with an erroneously low concentration.

4. The 5% CO_2 in the carbogen works with the $NaHCO_3$ as a physiological buffer. If water sits at room temperature for more than a day, then the amount of CO_2 will be low and the solution basic. Calcium phosphate will precipitate and will not readily go back into solution. Thus, the $CaCl_2$ should only be added to a well-bubbled solution.

5. If cerebellum slices are to be made, then it is advantageous to cut the skull beginning anteriorly and working backward.

6. We once tried slices from a brain that was kept in cold solution for an hour before slicing. The slices were fine.

7. Wide-mouth pipets seem to have fewer problems with sticking if they are coated with a layer of calcium deposits. Thus, we usually keep using the pipets rather than replacing them each day.

8. The temperature for this incubation varies somewhat, and some trial and error is necessary. For example, for studies of CA3, slices are often kept at 37°C for an hour. This allows the top layers of damaged cells to disintegrate. CA1 in these slices is dead.

References

1. Collingridge, G. L. (1995) The brain slice preparation: a tribute to the pioneer Henry McIlwain. *J. Neurosci. Methods* **59,** 5–9.
2. Blanton, M. G., LoTurco, J. J., and Krigstein, A. R. (1989) Whole cell recording from neurons in slices of reptilian and mammalian cerebral cortex. *J. Neurosci. Methods* **30,** 203–210.
3. Stuart, G. J., Dodt, H. U., and Sakmann, B. (1993) Patch-clamp recordings from the soma and dendrites of neurones in brain slices using infrared video microscopy. *Pflugers Arch.* **423,** 511–518.
4. Stuart, G. J. and Sakmann, B. (1994) Active propagation of somatic action potentials into neocortical pyramidal cell dendrites. *Nature* **367,** 69–72.
5. Colbert, C. M. and Pan, E. (2002) Ion channel properties underlying axonal action potential initiation in pyramidal neurons. *Nat. Neurosci.* **5,** 533–538.

11

Juxtacellular Labeling and Chemical Phenotyping of Extracellularly Recorded Neurons In Vivo

Glenn M. Toney and Lynette C. Daws

Summary

Extracellular recording of the action potential discharge of individual neurons has been an indispensable electrophysiological method for more than 50 yr. Although it requires relatively modest instrumentation, extracellular recording nevertheless provides critically important information concerning the patterning of intercellular communication in the nervous system. In 1996, Didier Pinault described "juxtacellular labeling" as "a novel and very effective single-cell labeling method" for revealing the morphology of extracellularly recorded neurons. Of particular interest for neuroscience is that juxtacellular labeling can be combined with immunocytochemistry and *in situ* hybridization histochemistry to reveal new and exciting information concerning the chemical phenotype of neurons whose electrophysiological properties have been characterized in vivo. By providing investigators with a means to "match" functional information from electrophysiological recordings with morphological and protein/gene expression data at the level of the single neuron, juxtacellular labeling has opened a new era in neuroscience research, one that holds the promise of an accelerated pace of discovery.

Key Words: Confocal microscopy; cellular morphology; electrophysiology; extracellular recording; histochemistry; immunocytochemistry; *in situ* hybridization.

1. Introduction

In the mammalian nervous system, neurons communicate (often over long distances) by generating patterns of action potentials. Ion channels lie at the heart of this process. Indeed, individual neurons in the intact nervous system express a host of ion channels with permeability and gating characteristics that determine the shape and patterning of action potentials. Recording the discharge of individual neurons has been at the forefront of neuroscience research

From: *Methods in Molecular Biology, vol. 337: Ion Channels: Methods and Protocols*
Edited by: J. D. Stockand and M. S. Shapiro © Humana Press Inc., Totowa, NJ

for many years, and this core methodology has been described at length else-where (1–3).

In its simplest form, extracellular single-unit recording involves introducing a glass, metal, or carbon fiber microelectrode with relatively low tip resistance (usually 5–40 MΩ) into the extracellular space adjacent to a neuron of interest. Action potentials of individual neurons are recorded as transient changes in electrode potential with the aid of a low-noise alternating current (AC)-coupled amplifier. Signals are referenced to an extracellularly located electrode com-prised of a nonpolarizing metal composite such as Ag/AgCl. In extracellular recording mode, voltages at the microelectrode are typically small (10–500 μV) owing to the low resistivity of extracellular fluid. Thus, considerable amplifi-cation is generally employed so that recorded signals effectively utilize the entire input voltage range of most commercially available analog-to-digital converters (ADCs; ±5 V). This is used in combination with a computer and data acquisition software to facilitate data storage, retrieval, and analysis.

The enduring utility of in vivo extracellular single-neuron recording can be attributed to the importance of determining the discharge response of neurons to specific afferent inputs and to the many accessory methods that have been developed over the years. For example, antidromic mapping and collision test-ing can be used to identify the location and branching pattern of the axon ter-minals of individual cells (4). Multibarrel glass electrodes can also be used to couple extracellular recording with pressure ejection or microiontophoresis of compounds such as receptor selective ligands (5). Such studies allow determi-nation of the contribution of specific neurotransmitters and receptors in con-trolling tonic activity and synaptically evoked neuronal responses. The behavior and organization of neural networks can be studied by simultaneously record-ing the discharge of multiple cells using an array of electrodes and sophisticated statistical correlation algorithms (6). Of particular interest is that the latter stud-ies can be achieved in conscious, freely behaving animals (7).

Here, we focus on a relatively new methodology termed juxtacellular label-ing, which allows histological staining and visualization of individual neurons whose discharge behavior has been electrophysiologically characterized. This methodology was developed by Didier Pinault and was introduced in 1996 (8). Anodal (positive) current pulses are passed through a glass recording electrode, which is filled with a solution containing biotinamide. The current pulses elec-trostatically repel positively charged biotinamide molecules, expelling them from the electrode. Because of the proximity of the electrode tip to the membrane of the recorded neuron, current pulses effectively cause single-cell electroporation-a transient opening created in the plasma membrane through which biotinamide molecules from the electrode gain entry into the cell interior, thus filling the cytoplasm of the soma and often the proximal dendrites (for review, see ref. 9).

When combined with antidromic activation, juxtacellular labeling allows the precise location and morphology of an individual neuron to be determined along with (in limited situations) its downstream connectivity. When further combined with immunocytochemistry or *in situ* hybridization histochemistry, the experimenter is able to determine the connectivity, morphology, and chemical phenotype of an individually recorded and electrophysiologically characterized neuron. Indeed, in vivo studies have used this experimental approach to record the discharge of individual neurons in brain regions including the thalamus *(8)*, medial septum *(10)*, dorsal raphe nucleus (DRN) *(11)*, rostral ventrolateral medulla *(9,12,13)*, cerebellum *(14)*, cochlear nucleus *(15)*, and retrotrapezoid nucleus *(16)* and, in the context of a single experiment, to determine the somatodendritic morphology of recorded cells along with details concerning their neurotransmitter content *(9,11,12)*, expression of receptor/transporter protein/messenger (m)RNA *(9,10)*, cytosolic enzymes *(9,12)*, and ion channel subunit mRNA *(13)*.

2. Materials

1. Borosilicate glass (1.5 mm od, 0.86 mm id, World Precision Instruments, Inc., Sarasota, FL).
2. Narishige vertical puller (model PE-2, Narishige International USA, Inc., East Meadow, NY).
3. Axoclamp 2B amplifier with HS-2A x0.1LU headstage (Axon Instruments, Inc., Foster City, CA).
4. AC preamplifier (Grass, model P15D, Astromed, Inc., Grass Instruments Division, West Warwick, RI).
5. ADC (model 1401plus/micro1401, Cambridge Electronic Design, Inc., Cambridge, UK).
6. Window discriminator (model 121, World Precision Instruments).
7. Data acquisition/analysis software (*Spike2*, v4.31, Cambridge Electronic Design, Inc.).
8. Pulse generator (model PulseMaster, World Precision Instruments).
9. Sodium acetate, sodium chloride (Sigma, St. Louis, MO).
10. Biotinamide (Molecular Probes, Eugene, OR) or neurobiotin (*N*-[2 aminoethyl] biotinamide HCl, Vector Laboratories, Burlingame, CA).
11. Anesthesia for rats: cocktail containing urethane (1.0 g/mL) and α-chloralose (0.1 g/mL) at a dose of 800 mg/kg urethane and 80 mg/kg chloralose ip. (Combine 1 g urethane with 0.1 mL normal saline and warm to approx 40°C, then add chloralose and vortex/stir until dissolved.)
12. 0.1 *M* Phosphate-buffered saline (PBS).
13. PBS containing 4% paraformaldehyde (PFA).
14. PBS supplemented with 30% sucrose.
15. Freezing microtome/cryostat.
16. Polyvinylpyrrolidine (PVP-40) cryoprotectant: 30 g sucrose, 30 mL ethylene glycol, and 1.0 g PVP-40 brought to 100 mL with PBS.

17. 100 m*M* Tris-buffered saline (TBS).
18. Triton X-100.
19. Streptavidin-Texas red or streptavidin-FITC (fluorescein isothiocyanate) (Jackson ImmunoResearch Laboratories, West Grove, PA).
20. Avidin-peroxidase conjugate (ABC Vectastain Kit, Vector).
21. 3,3'-Diaminobenzidine tetrahydrochloride (DAB), hydrogen peroxide, and nickel sulfate hexahydrate (Sigma).
22. Normal goat serum (NGS; unless primary antibody is from goat, in which case use donkey serum).
23. Primary antibody of interest (e.g., monoclonal antitryptophan hydroxylase antibody). For immunohistochemistry, this antibody may be conjugated to a fluorophore, such as FITC or Texas red. Alternatively, an appropriate secondary antibody conjugated to FITC (Caltag Laboratories, Burlingame, CA, cat. no. M32601) or Texas red may be used to identify the primary antibody.
24. Superfrost Plus slides (Fisher, Hampton, NH).
25. Ethanol (70, 95, and 100%).
26. Xylene.
27. Cytoseal 60 (Fisher).

3. Methods
3.1. In Vivo Extracellular Single-Unit Recording

1. Borosilicate glass microelectrodes are pulled to a fine-tip diameter (1–2 μm) using a Narishige vertical puller. The tips are blunted by making contact with the broken flat surface of a glass rod positioned under a microscope. The final tip resistance is typically 15–40 MΩ.
2. Microelectrodes are filled with 0.5 *M* sodium acetate supplemented with 5% biotinamide.
3. We obtain recordings with a direct current (DC) intracellular amplifier. We prefer the Axoclamp 2B in bridge mode (Axon). We then pass the ×10 output of the DC amplifier to a battery-powered model P15D AC preamplifier (Astro-Med). The amplifier has half-amplitude frequency filters that we usually set to a bandpass of 0.3–3.0 kHz. Use of a 60-Hz notch filter can also be advantageous.
4. The output of the AC amplifier is then sent to an ADC (model 1401plus, Cambridge Electronic Design) and to a window discriminator (model 121, World Precision Instruments). The window discriminator multiplex output is sent to an analog oscilloscope and to the ADC to monitor cell discharge. We use *Spike2* (v4.31) data acquisition and analysis software (Cambridge Electronic Design). Data is stored on a computer.

3.2. Juxtacellular Labeling

1. The juxtacellular labeling procedure has been described elsewhere *(8,9)*. Once a single-unit recording is obtained, as evident by spikes of uniform shape and

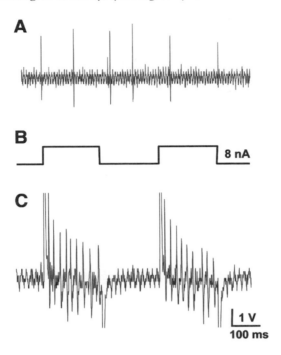

Fig. 1. (**A**) Single-unit recording of the spontaneous discharge of a slow-firing (0.5-Hz) neuron in the DRN of a C57Bl/6 mouse. (**B**) Anodal (positive) current pulses (200-ms 50% duty cycle) were passed through the recording electrode, and the amplitude was adjusted to 8 nA. (**C**) The discharge of the recorded neuron became entrained to the timing of current pulses. Note that current-switching artifacts are clipped to emphasis the entrained discharge. Voltage calibration is postamplification.

    amplitude, and physiological studies have been completed, juxtacellular labeling is initiated (*see* **Fig. 1A**).

2. The Axoclamp 2B amplifier has convenient front-mounted step command switches that facilitate juxtacellular labeling by allowing accurate and rapid adjustment of the amplitude of current pulses delivered through the recording electrode (*see* **step 3**). When using the standard HS-2A x0.1LU head stage, the full-scale output is ±19.99 nA, which is typically more than adequate for juxtacellular labeling (*see* **Note 1**).

3. Anodal (positive) current pulses are delivered through the recording electrode. The standard approach is to deliver 200-ms pulses with a 50% duty cycle (*see* **Fig. 1B**). It is most convenient to trigger current pulses with an external pulse generator such as the PulseMaster (World Precision Instruments). Alternatively, the digital-to-analog output available with most digital data acquisitions systems can be programmed for the same purpose.

4. The amplitude of current pulses is gradually increased while continuously monitoring the discharge of the individually recorded neuron. The goal is to entrain the discharge of the recorded neuron to the timing of current pulses (*see* **Fig. 1C**). The amplitude of current pulses required to entrain discharge varies. Factors likely to influence this include specific properties of the recorded neurons as well as properties of the recording electrode, such as tip resistance, diameter, and overall shape.

5. In our experience, entrainment of discharge among neurons of the DRN is successful in the majority of cases at current amplitudes ranging from 2 to 8 nA. Occasionally, entrainment may require that the recording electrode be moved in small increments toward the recorded neuron. Depending on properties of the recorded cell, an increase in signal noise may occur as an indication that the microelectrode has reached the "juxtacellular" location. In this position, current pulses of the appropriate amplitude will most often result in discharge entrainment. During entrainment, electroporation of the cell membrane will result in filling of the cytoplasm with biotinamide. It is sometimes observed that cells that resist entrainment will succumb if the amplitude of current pulses is increased to 15–20 nA for a few cycles and then quickly reduced (*see* **Note 2**).

6. After discharge becomes entrained to the timing of current pulses, successful filling will typically be achieved within a short time, usually 30 s to 5–10 min. How long entrainment must be maintained for adequate filling to occur depends on several factors, including the size and morphology of the cell, electrode shape/resistance, and concentration of biotinamide in the recording electrode. Another factor that may influence the efficacy of entrainment and cell filling is the nature of the microelectrode filling solution. Some investigators report improved results when electrodes are filled with 0.5 M sodium acetate compared to sodium chloride. Presumably, this reflects the valence or mobility of the counteranion in solution with biotinamide.

7. Once entrainment has been maintained for a sufficiently long period, the amplitude of current pulses is reduced, and pulse delivery is terminated. It is advisable to record spontaneous cell discharge for several minutes thereafter to assess the "health" of the recorded neuron following the entrainment/labeling procedure. At this point, the recording electrode can be removed. Depending on the goal of the experiment, it may be advisable to delay perfusion fixation of the animal for up to several hours to maximize diffusion/distribution of biotinamide throughout the cell interior. The last is particularly important if the morphology of dendrites is to be examined and if local connectivity is to be observed.

3.3. Recovering Biotinamide-Filled Neurons

1. After recording and juxtacellularly labeling a neuron (*see* **Subheading 3.2.**), the still-anesthetized rat is perfused transcardially with 100 mL PBS followed by perfusion with 330 mL chilled (4°C) PBS containing 4% PFA to fix the brain tissue. The rate of perfusion is approx 33 mL/min. For fixation of mouse brain, the volume and rate of transcardiac perfusion are reduced to 10% of the values for the rat.

2. The brain is removed and postfixed in PFA for 4–24 h. To ensure thorough cryoprotection, brains are then transferred to 30% sucrose-PBS until the tissue block sinks (~48 h). At this time, tissue can either be placed in frozen storage or cut into sections 30-µm thick. If stored, brains should be frozen in chilled isopentane before placing them in a freezer. Isopentane is placed in a glass beaker surrounded by crushed dry ice until chilled (~15 min), and brains are then immersed until fully frozen. We wrap brains in aluminum foil and place them directly in a freezer at −80°C. If brains are immediately sectioned, a freezing microtome/cryostat is used at −20°C. After cutting, sections can be stored at −20°C in PVP-40 cryoprotectant prior to processing for recovery of juxtacellular labeling and immunostaining.

3. Tissue sections stored in cryoprotectant at −20°C are warmed to room temperature (30 min) and rinsed (five times for 5 min/rinse) in TBS containing 0.1% Triton X-100.

4. Biotinamide-filled cells are recovered by reacting tissue with streptavidin in TBS containing 0.1% Triton X-100. Streptavidin can be conjugated to a wide variety of fluorochromes. We have used streptavidin-Texas red or streptavidin-FITC at a concentration of 1:200. Alternatively, soma and dendritic morphology can be analyzed by recovering biotinamide-filled cells using a standard avidin-peroxidase/diaminobenzadine reaction with metal (Ni^{2+}) intensification. The latter has the decided advantage that the resultant staining is blue/black and is effectively permanent.

5. Sections are incubated in TBS containing streptavidin-Texas red (1:200) for 1 h at room temperature and then for 24 h at 4°C with mild agitation on a rocker table. We carry out the reaction in a 1.5-mL microcentrifuge tube to minimize the amount of fluorochrome used. Sections are covered with foil to minimize photobleaching.

6. Sections are warmed to room temperature and rinsed in TBS (five times for 5 min/rinse), mounted on Superfrost Plus slides (Fisher), and air-dried overnight while covered with aluminum foil.

7. Sections are dehydrated in a series of ethanol (3 min each in 70, 95, 95, 100%), cleared (defatted) in xylene (2–3 min each), and cover slipped with Cytoseal 60 (Fisher) (*see* **Fig. 2**).

3.4. Colocalization of Cellular Proteins by Immunocytochemistry

1. Immunocytochemical staining to colocalize immunoreactive proteins within juxtacellularly labeled neurons can be performed before or after recovery of biotinamide-filled cells. The optimal sequence of staining usually depends on whether juxtacellularly labeled cells are recovered using a fluorescent marker or a peroxidase method as previously described. If different fluorochromes are used for recovering the juxtacellularly labeled cell and for immunostaining, the sequence used for the two staining steps will usually depend on their relative intensities. The intensity of fluorescence will depend on a number of factors, including the degree of biotinamide filling (juxtacellular label), the density of

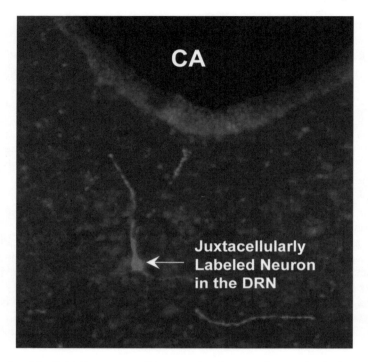

Fig. 2. A wide-field fluorescent (Texas red) micrograph reveals the juxtacellularly labeled neuron in the DRN for which the entrainment may be seen in **Fig. 1**. Note that the soma and proximal dendrites are filled with biotinamide. CA, cerebral aqueduct.

    protein expression, antibody affinities, and the type of amplification system used for staining.
2. As noted in the **Subheading 1.**, studies to date have successfully combined juxtacellular labeling with immunocytochemical staining to localize a number of cellular proteins within electrophysiologically characterized central neurons. Here, we present our methodology for immunostaining for tryptophan hydroxylase (TPH), the rate-limiting enzyme in the synthesis of serotonin (5HT). We have used this approach to neurochemically phenotype serotonergic neurons in the DRN of C57BL/6 mice (*see* **Fig. 3**).
3. First, tissue sections through the DRN (stored in PPV-40 cryoprotectant) are warmed to room temperature (30 min) and placed in chambers with a mesh bottom to facilitate transfer to various solutions in the staining protocol.
4. Sections are rinsed in TBS (five times for 5 min/rinse) and incubated in a monoclonal primary anti-TPH antibody (Sigma, cat. no. T0678) diluted 1:500 in TBS containing 1% NGS and 0.1% Triton X-100 (*see* **Note 3**).
5. Sections are incubated with gentle agitation at room temperature for 1 h and at 4°C for 48 h.

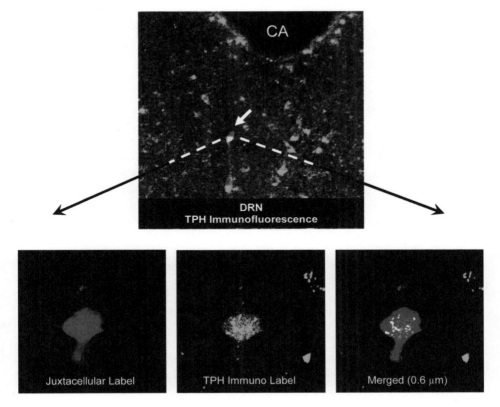

Fig. 3. **Top,** laser scanning confocal photomicrograph (×10) showing colocalization of tryptophan hydroxylase (TPH) immunofluorescence (green) in a juxtacellularly labeled neuron (red) in the DRN of a C57Bl/6 mouse. Arrow indicates the recorded neuron. **Bottom,** high-magnification (×60) image of juxtacellularly labeled DRN neuron (left), TPH immunofluorescence (center), and a merged image showing colocalization at a 0.6-μm z-plane resolution. CA, cerebral aqueduct.

6. Next, tissue sections are again warmed to room temperature and rinsed in TBS (five times for 5 min/rinse).
7. Sections are then incubated with gentle agitation in goat antimouse immunoglobulin G3 secondary antibody conjugated to FITC (Caltag Laboratories, cat. no. M32601) for 1 h at room temperature and 6–9 h at 4°C. Sections are covered with aluminum foil to minimize photobleaching.
8. If sections were processed earlier for recovery of juxtacellularly labeled cells, the procedure is now complete, and sections can be rinsed in TBS (five times for 5 min/rinse), cleared, and cover slipped as indicated in **Subheading 3.3.4.**
9. If sections have not yet been processed for recovery of the juxtacellular label, the procedure may now be performed by immersing sections in streptavidin-Texas

red diluted 1:200 in TBS containing 0.1% Triton X-100 and 1% NGS and incubating sections for 1 h at room temperature and for 24 h at 4°C.

10. Sections are then rinsed in TBS, cleared, mounted on slides, and cover slipped as previously described.

11. Sections can then be viewed under an appropriately equipped fluorescent microscope (wide field or confocal) to determine the presence of double-labeled neurons (*see* **Fig. 3**).

4. Notes

1. We have successfully implemented juxtacellular labeling with a variety of instruments. It is possible to use a relatively inexpensive DC amplifier like the model 767 Electrometer from World Precision Instruments; it has a unitary gain headstage. Because this unit also has only 10× and 50× output, delivery of even small current pulses (<10 nA) will saturate our AC amplifier. To avoid this, we impose an attenuating (×0.1) amplifier/isolation unit (model SIU5A, Grass Instruments) before the signal is passed to the AC amplifier. It is a simple matter then to use a pulse generator to drive the external input to the current isolator circuit of the amplifier to trigger and grade the amplitude of current pulses through the recording microelectrode. The amplitude of current pulses can be readily monitored by simply calibrating a channel of the computer data acquisition software or an oscilloscope.

2. Labeling multiple neurons: it is possible for the juxtacellular labeling method to result in filling of more than one cell. The incidence of this is generally low, particularly when care is taken to monitor cell discharge continuously to ensure that only the individually recorded neuron becomes entrained by current pulses. The frequency of labeling multiple neurons in the DRN of mice in our experience is low (<5%). Guyenet et al. *(9)* also reported that multiple-cell filling is infrequent in the rostral ventrolateral medulla. As these investigators noted, however, the frequency of labeling multiple cells may increase in brain regions where the packing density of neurons is high.

3. High background staining: the TPH immunocytochemical staining procedure described for mice may be improved because of more recent availability of antimouse primary antibodies raised in nonmurine species. Alternatively, immunostaining for serotonin may be performed with excellent results.

Acknowledgments

We thank Mr. Alfredo Calderon for excellent technical assistance. This work was supported by National Institute of Health grants HL71645, HL76312 (G. M. T.), and MH64489 (L. C. D.).

References

1. Aston-Jones, G. S. and Siggins, G. R. (1995) Electrophysiology, in *Psychopharmacology: The Fourth Generation of Progress* (Bloom, F. E. and Kupfer, D. J., eds.), Raven Press, New York, pp. 41–63.

2. Sherman-Gold, R. (ed.). (1993) Instrumentation for measuring bioelectric signals from cells, in *The Axon Guide For Electrophysiology and Biophysics Labortory Techniques*, Axon Instruments, Foster City, CA, pp. 17–24.
3. Sherman-Gold, R. (ed.) (1993) Same ref. below, Axon Instruments, Foster City, CA, pp. 25–80.
4. Lipski, J. (1981) Antidromic activation of neurones as an analytic tool in the study of the central nervous system. *J. Neurosci. Methods* **4,** 1–32.
5. Stone, T. W. (1985) *Microiontophoresis and Pressure Ejection*, Wiley, New York.
6. Blanche, T. J., Spacek, M. A., Hetke, J. F., and Swindale, N. V. (2004) Polytrodes: high density silicon electrode arrays for large scale multiunit recording [Epub]. *J. Neurophysiol.* PMID 15548620.
7. Super, H. and Roelfsema, P. R. (2005) Chronic multiunit recordings in behaving animals: advantages and limitations. *Prog. Brain Res.* **147,** 263–282.
8. Pinault, D. (1996) A novel single-cell staining procedure performed in vivo under electrophysiological control: morpho-functional features of juxtacellularly labeled thalamic cells and other central neurons with biocytin or neurobiotin. *J. Neurosci. Methods* **65,** 113–136.
9. Guyenet, P. G., Stornetta, R. L., Weston, M. C., McQuiston, T., and Simmons, J. R. (2004) Detection of amino acid and peptide transmitters in physiologically identified brainstem cardiorespiratory neurons. *Auton. Neurosci.* **114,** 1–10.
10. Bassant, M. H., Simon, A., Poindessous-Jazat, F., Csaba, Z., Epelbaum, J., and Dournaud, P. (2005) Medial septal GABAergic neurons express the somatostatin sst_{2A} receptor: functional consequences on unit firing and hippocampal theta. *J. Neurosci.* **25,** 2032–2041.
11. Allers, K. A. and Sharp, T. (2003) Neurochemical and anatomical identification of fast- and slow-firing neurons in the rat dorsal raphe nucleus using juxtacellular labeling methods in vivo. *Neuroscience* **122,** 193–204.
12. Schreihofer, A. M. and Guyenet, P. G. (1997) Identification of C1 presympathetic neurons in rat rostral ventrolateral medulla by juxtacellular labeling in vivo. *J. Comp. Neurol.* **387,** 524–536.
13. Washburn, C. P., Bayliss, D. A., and Guyenet, P. G. (2003) Cardiorespiratory neurons of the rat ventrolateral medulla contain TASK-1 and TASK-3 channel mRNA. *Resp. Physiol. Neurobiol.* **138,** 19–35.
14. Simpson, J. I., Hulscher, H. C., Sabel-Goedknegt, E., and Ruigrok, T. J. (2004) Between in and out: linking morphology and physiology of cerebellar cortical interneurons. *Prog. Brain Res.* **148,** 329–340.
15. Arnott, R. H., Wallace, M. N., Shackleton, T. M., and Palmer, A. R. (2004) Onset neurones in the anteroventral cochlear nucleus project to the dorsal cochlear nucleus. *J. Assoc. Res. Otolaryngol.* **5,** 153–170.
16. Mulkey, D. K., Stornetta, R. L., Weston, M. C., et al. (2004) Respiratory control by ventral surface chemoreceptor neurons in rats. *Nat. Neurosci.* **7,** 1360–1369.

12

Carbon Fiber Amperometry in the Study of Ion Channels and Secretion

Duk-Su Koh

Summary

Activation of Ca^{2+} channels in the plasma membrane or on internal Ca^{2+} stores raises cytosolic Ca^{2+} concentration ($[Ca^{2+}]_c$). Among diverse functions of Ca^{2+} signals, the induction of exocytosis—the process in which the contents of secretory vesicles are released by their fusion to the plasma membrane—is one of the most important. For example, in neurons and endocrine cells, it allows intercellular communication and secretion of biomolecules. Exocytosis can be detected by several physical and chemical means. By chemically oxidizing the released secretory products at a fixed electrode potential, carbon fiber amperometry provides excellent temporal and spatial resolution in detecting exocytosis. This method, together with other biophysical techniques such as patch clamp and Ca^{2+} microphotometry, has greatly contributed to our understanding of the molecular mechanisms involved in the stimulus-secretion coupling. However, amperometry can be performed only on cells that secrete oxidizable molecules. To overcome this limit, we have developed a protocol of loading cells with exogenous neurotransmitters that readily oxidize on a carbon electrode. Several cell types have been successfully loaded, exocytosis of secretory vesicles has been demonstrated, and in pancreatic duct epithelial cells, the modulatory signals of exocytosis have been studied in detail.

Key Words: Amperometry; Ca^{2+} channels; carbon fiber electrode; dopamine; exocytosis; exogenous loading; oxidation; secretion; serotonin.

1. Introduction

Basal Ca^{2+} concentration in the cytoplasm is normally kept as low as approx 100 nM or less, and it increases when Ca^{2+}-permeable ion channels in the plasma membrane or in the intracellular Ca^{2+} stores become active. In addition to their influence on the membrane potential as charge carriers, Ca^{2+} ions orchestrate

From: *Methods in Molecular Biology, vol. 337: Ion Channels: Methods and Protocols*
Edited by: J. D. Stockand and M. S. Shapiro © Humana Press Inc., Totowa, NJ

multiple effects, including muscle contraction, gene expression, and activation of Ca^{2+}-dependent proteins as well as secretion of diverse biomolecules (1,2). All eukaryotic cells have vesicular traffic systems to insert membrane proteins and to package proteins for secretion. Exocytosis, the final event of the secretory pathway, is the fusion of secretory vesicles to the plasma membrane. Some vesicles exocytose constitutively, whereas others are regulated by intracellular signals, of which Ca^{2+} is the most common.

The secretory products from multicellular preparations can be measured by a variety of sensitive biochemical methods, such as radioimmunoassay. Measurement of exocytosis at the single-cell level has become possible using relatively recent biophysical methods such as capacitance measurement (reviewed in ref. 3) and optical techniques such as labeling the secretory vesicles with FM 1-43 fluorescent dye (reviewed in ref. 4). Those methods allow detection of the vesicle dynamics with subsecond time resolution.

Fusion of even a single vesicle and release of its contents can be directly studied by electrochemical techniques using the carbon fiber electrode (CFE) (5). This method detects secreted molecules based on their oxidation (or, less frequently, reduction). If the electrode potential sufficiently exceeds the chemical potential for oxidation of a molecule, then the molecule oxidizes and delivers electrons to the electrode, resulting in an oxidation current.

The electrode potential can be varied (voltammetry) or fixed (amperometry). The former approach allows the partial identification of the released molecule; the latter monitors the time-course of exocytosis continuously at submillisecond resolution but without information of the released molecule's chemical nature (6,7). If the chemical nature of the secreted molecule is known, then amperometry has been the method of choice for measuring exocytosis from isolated single cells. Amperometry is sensitive enough to detect the cargo from single vesicles (quantal release) and stable enough for the recording of exocytosis for an extended time period (see Fig. 1). This stability is because of its noninvasive nature, distinct from whole-cell capacitance measurements. Most advantageously, it detects exocytotic events without confounding by concurrent endocytosis. Application of small electrodes with a tip size less than 2 μm also revealed the hot spots of exocytosis (8). Amperometry has been successfully applied to neuronal (9,10), endocrine (5,11), and some nonexcitable cells (12–14). In the first two cell types, endogenous neurotransmitters such as serotonin, dopamine, norepinephrine, or epinephrine were detected. In the case of nonexcitable cells that do not secrete the oxidizable neurotransmitters, exogenous oxidizable molecules were loaded into the cells.

In this chapter, I provide a practical guide to applying carbon fiber amperometry to single cells and a method for loading artificial reporters. Theoretical background and a variety of related methods have been reviewed in detail previously *(6,7)*.

2. Materials

2.1. Preparation of Carbon Fiber Electrodes

1. Carbon fiber (T650 for 5.1 μm or P25 for 11 μm; BP Amoco Polymers, Alpharetta, GA).
2. Disposable 10-μL micropipet tip (Rainin, Oakland, CA).
3. Ethanol (reagent quality; Sigma, St. Louis, MO).
4. Plastic 60-mm culture dish with a silicone elastomer (Sylgard184, Dow Corning, Midland, MI) layer on the bottom.
5. Fine forceps (no. 5, Fine Science Tools, Vancouver, Canada) with or without both tips covered with polyethylene (PE) tubing (Intramedic PE 50, BD Inc., Franklin Lakes, NJ).
6. Home-made puller/cutter (described in **Subheading 3.**).
7. Dissecting microscope (×5–40) with a fiber-optic illuminator.

2.2. Amperometric Measurements

1. Cultured cells plated on a small glass chip. Subconfluent cell density is optimal so that single cells isolated from their neighbors can be found. For the recording shown in **Fig. 1**, we used a cultured pancreatic duct epithelial cell 2 d after plating *(14)*.
2. External solution such as the Na^+-rich saline solution used in our experiments (140 mM NaCl, 2.5 mM KCl, 2 mM CaCl$_2$, 1 mM MgCl$_2$, 10 mM D-glucose, and 10 mM HEPES at pH 7.3 adjusted with NaOH).
3. CFEs with a tip size of 5 or 11 μm.
4. Ag/AgCl reference electrode. A 0.5-mm thick silver wire is scraped with fine sandpaper (150 grit) and cleaned with absolute ethanol. It is then immersed in a commercial Clorox bleach solution for about 30 min until the surface appears dark brown. An alternative way is to apply 1.5 V between the reference electrode (positive) and another silver wire (negative) in 0.1 N HCl solution for a few minutes.
5. Amplifier: patch clamp amplifiers, such as EPC-9 or -10 (HEKA Elektronik, Lambrecht, Germany) are suitable for measuring picoampere-range amperometric currents. Other amplifiers designed for electrochemistry, like Chem-Clamp (Dagan, Minneapolis, MN), can also be used.
6. Analog-to-digital (A/D) converter: EPC-9 and -10 have internal A/D converters. With other types of amplifiers, an additional A/D converter is needed to digitize the current signals for computer storage (*see* **Note 1**).
7. Manipulator: motorized, piezo-driven, or hydraulic manipulators are suitable for positioning the tip of a CFE near the cell. For lengthy recording (>10 min), mechanical stability with minimal drift is critical so that the same membrane area is monitored during the entire experiment.

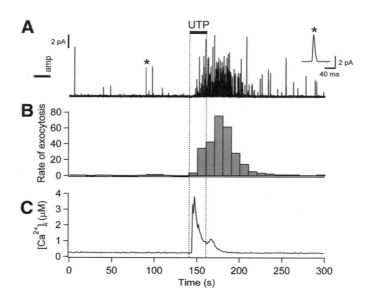

Fig. 1. Simultaneous measurement of amperometric signal and $[Ca^{2+}]_c$ in a pancreatic duct epithelial cell. The cell was pretreated with 1 μM indo-1 AM for 30 min and
then with 70 mM dopamine for the next 40 min at room temperature. (**A**) Exocytotic
events monitored using an 11-μm carbon fiber electrode. After recording the basal
level of exocytosis, the cell was treated with 10 μM UTP to activate the P2Y$_2$ purinergic receptors expressed in the cell. Inset: A quantal event marked with an asterisk is
illustrated at an expanded time scale. Electrical charge acquired by integrating the single
amperometric current indicates that the vesicle contained about 110,000 dopamine
molecules. Filter and sample frequency were 100 and 500 Hz, respectively. Na$^+$-rich
saline solutions were perfused near the cell using a local multibarrel system (solution
exchange time less than 0.5 s *[22]*). (**B**) Histogram for the rate of exocytosis as measured events per 10-s time bin. (**C**) Cytosolic Ca^{2+} level in the same cell was monitored
using the Ca^{2+}-sensitive indo-1 dye every 1 s (for details, refer to **ref.** *14*). Phospholipase C and IP$_3$ signaling linked to P2Y$_2$ receptor induced the rise of Ca^{2+} and Ca^{2+}-
dependent exocytosis in this nonexcitable cell as detected in **A**. Comparison of the rate
of exocytosis (**B**) and $[Ca^{2+}]_c$ (**C**) reveals that exocytosis shows slower on- and offset
reactions compared to Ca^{2+} rise.

2.3. Loading of Exogenous Oxidizable Molecules Into Secretory Vesicles

1. Dopamine, serotonin, and ascorbic acid (reagent grade; Sigma).
2. For loading by specific transporters expressed in the cell type: two 100 mM stock
   solutions containing either the neurotransmitters or ascorbic acid are dissolved in
   the culture medium. Aliquots of approx 100 μL stock solution can be frozen and
   stored for at least 2–5 d.

3. For loading by diffusion or by endocytosis: freshly prepared 2 mL loading solution similar to the Na^+-rich external recording solution, *except* a high concentration of serotonin or dopamine (30–70 mM) equimolarly replaces NaCl. To accelerate the preparation of the loading solution, the following recipe is useful: add an appropriate amount of catecholamines (19 mg for dopamine) to 1.43 mL of an external solution containing 70 mM NaCl (pH 7.3). This produces a loading solution containing 70 mM dopamine and pH becomes around 7.0 because of the added dopamine (we do not adjust the pH). This solution can be used for loading or mixed with an appropriate amount of normal Na^+-rich external solution when a loading solution containing fewer catecholamines is needed. Finally, add ascorbic acid (final concentration 1–2 mM) to the loading solution to reduce spontaneous oxidation of dopamine during the loading process.

2.4. Data Analysis

For software that can detect spikes and integrate current traces to analyze amperometric recordings, we use *Igor* software (WaveMetrics, Lake Oswego, OR) with a macro written for the spike detection. Certain algorithms for the detection of synaptic miniature events can be used for the purpose (e.g., *pClamp9*, Axon Instruments, Union City, CA). Peak detection algorithms typically calculate a mean current within a time window (a set of N consecutive data points) and systematically move the window along the data. When the difference of the current averages of two consecutive windows is larger than a threshold, the second one is regarded as the rising phase of an amperometric spike. The rule of thumb suggests that the threshold of detection is five times larger than the standard deviation of baseline noise *(15)*. This criterion rules out the possibility of misjudging a noise event as a true amperometric signal.

3. Methods

3.1. Preparation of CFEs

CFEs can be manufactured in a variety of forms. The basic requirement is that most carbon surface *except* the sensing area (typically the tip of the electrode that contacts the cell) is shielded. Shielding materials include glass *(7)*, nonconducting electrodeposited paint *(16)*, PE *(6)*, and polypropylene *(11)*. Tight annealing of the carbon fiber surface with the shielding material appears critical in reducing the capacitance of the electrode and the consequent electrical noise. Inspection of the electrodes using scanning electron microscopy has revealed that polypropylene is superior to glass in this regard (*[17]*; *see* **Fig. 2B**).

I describe CFEs pulled from disposable 10- mL polypropylene micropipet tips, as originally developed by Zhou and Misler *(11)*. This type of CFE is easily produced in the laboratory with a relatively simple puller, and it is also commercially available (ProCFE, Dagan). A semiautomatic puller and a cutting

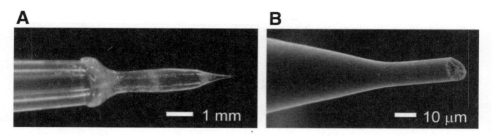

Fig. 2. Polypropylene (PP)-coated carbon fiber electrodes inspected under dissecting microscope (**A**) or scanning electron microscope (**B**). (**A**) Shape of PP insulation is determined by pull speed. Thick PP insulation at electrode tips provides good mechanical stability. (**B**) This high magnification of the tip demonstrates smooth annealing of the PP coating on carbon fiber. When inspected at higher magnification, even the end piece of the carbon fiber is partially covered by patches of PP layer. Optimal length of carbon fiber at the electrode tip should be this long or shorter.

device for polypropylene-coated electrodes were previously described *(17)*. Here, a simpler manual puller with easier construction is described. The essential elements of this puller consist of a heater and a translational stage (**Fig. 3**); other elements are explained in **Subheading 3.1.1.**, **steps 4–6** and the legend to **Fig. 3**.

Manufacture of a CFE consists of two steps: encasing a carbon fiber in a micropipet tip (*pulling*) and then trimming the electrode tip (*cutting*). Because melting a plastic tube with a carbon fiber inside is unpredictable, the whole pulling process is performed under a microscope (×20–40).

3.1.1. Insulation of Carbon Fiber With Plastic Coating

1. A bundle of 11-μm carbon fiber is cut 1.5–2 cm long and maintained in 100% ethanol in a 60-mm plastic culture dish (*see* **Note 2**).
2. Dip a micropipet tip into the ethanol, and a carbon fiber is threaded into the tip of a micropipet using a fine forceps (no. 5; *see* **Note 3**). Bright side illumination using a fiber-optic illuminator increases the visibility of carbon fibers during handling in the ethanol. The micropipet tip fills with ethanol by the capillary effect.
3. After insertion of the carbon fiber, ethanol remaining in the micropipet is eliminated by touching the tip with a filter paper or soft tissue. When dried, the carbon fiber will stay inside the micropipet, presumably because of electrostatic force (*see* **Note 4**).
4. Both ends of the micropipet are held on the pulling stage; the wide-opening end of the pipet is fixed using a tapered adaptor (adaptor A) attached to the translational stage (**Fig. 3**). The pipet tip is inserted through the heating coil and fixed inside a metal tube originating from a syringe needle (16 gage, the end piece of

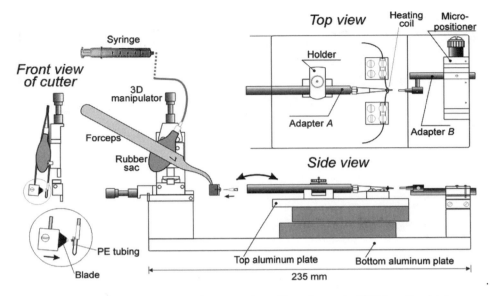

Fig. 3. Proposed configuration for the carbon fiber electrode (CFE) puller (right part in the side view) and the cutter (left part). To build the puller, a one-dimensional translation stage is mounted on the bottom aluminum plate by screws. Heating coil, holder, and adaptor A for micropipet tip are installed on the top aluminum plate, which is fixed to the stage by screws. *See* text for the pulling operations (**Subheading 3.3.1.**). For cutting the carbon fiber at the CFE tip, the direction of adaptor A with a pulled CFE changes. The cutter trims the carbon fiber using a piece of razor blade fixed on a tip of a forceps (front view; cutting elements are expanded ×2 for clearer visibility). The carbon fiber is positioned on the PE tubing (marked with an asterisk in the expanded view) that is attached to the other side of the forceps. The PE tubing side of the forceps is firmly fixed to the manipulator by a screw (illustrated here but actually invisible from the front). Refer to text for the cutting process. Drawn to scale except for 5-mL glass syringe.

adaptor B). Adaptor A is held relatively loosely so that it can slide within the holder when forced by hand during installation of a pipet tip, but it provides a stretching force on the molten plastic during pulling. Adaptor B can be aligned to the center of the micropipet tip in the Z direction with the aid of a swinging arm, in the Y direction with a micropositioner, and in the X direction by adjusting the shaft through a hole. A screw tightens the position of adaptor B after alignment.

5. Heating the coil (tungsten wire, 0.14 Ω/cm) with 5–12 V direct or alternating current for several seconds melts the micropipet tip, and the molten plastic separates to both sides, leaving a long, exposed carbon fiber between them. Then, stop heating by turning off the current.

6. Immediately after heating is stopped, the left side of the pipet (wide opening) is slowly pulled by the manual shift of the translational stage while the molten plastic gradually solidifies. The right side of the pipet remains stationary. At this stage, the two sides of micropipet are still connected by the carbon fiber. Touching the tip end of the fiber with a forceps without PE protection yields a CFE with a long carbon fiber hanging from the electrode tip.
7. CFEs with a smaller tip diameter (e.g., 5 µm; T650 carbon fiber) can be produced using the same techniques.

3.1.2. Trimming the Electrode Tip

The next step of CFE fabrication is to cut the carbon fiber at the point where the insulation ends. This can be done in several ways: the easiest is by hand using a sharp blade or an iris scissors (*6,11*). To reduce the effect of hand tremor, the cutting devices can be mounted in a fixture, preferably on a manipulator or a translational stage (Z. Zhou, personal communication, Nov. 1996). We have designed a few different cutters to promote stability and control. A cutting device using a miniature solenoid actuator was described previously (*17*).

Here, I describe a simpler cutter that can be constructed easily in a laboratory without the support of a fine mechanical machine shop (**Fig. 3**, left). It uses a hydraulic remote control to press a sharp razor blade against the carbon fiber. The left tip of a forceps has a holder for a piece of a razor blade, and the right tip adapts a supporting base for carbon fiber (a piece of Intramedic PE 50 tubing). These elements are firmly attached to the forceps tips by soldering. The piece of the razor blade is cut using a pair of scissors and ground using a stone grinder to have a narrow cutting edge (**Fig. 3**, front view). The blade and the PE tubing are aligned by adjusting and bending the metal wire holding the PE tubing. The position of the razor blade is controlled by a hydraulic device consisting of a rubber sac, a syringe, and a forceps (no. 7 bent-tip shape, Fine Science Tools).

A rubber bulb used for Pasteur pipets is fixed to the inside of the forceps using epoxy glue (*see* **Note 5**). The rubber bulb is connected to a 5-mL glass syringe filled with mineral oil via a connector and a PE tubing (1.5-mm i.d.). A three-way stopcock attached to the syringe is useful for removing air in the system or refilling the mineral oil. Finally, the cutting forceps is fixed on a three-dimensional manipulator through a hole drilled on the side where the stationary support PE tubing is fixed (**Fig. 3**, front view). Trimming steps are as follows:

1. Install a pulled CFE on adaptor A, which is now directed toward the cutter. Then, insert the extending carbon fiber at the electrode tip between the blade and the plastic tube.
2. Position the exact point where the carbon fiber is to be cut in the center of the PE tubing using the three-dimensional manipulator.
3. Advance the blade by retracting the mineral oil into the syringe to cut the carbon fiber.

4. The electrode tip is inspected under a microscope (×400) to check for sharp edges that could damage the cell in case of direct contact during measurement (*see* **Note 6**).

3.2. Amperometric Measurements

1. Fill a CFE with either 3–4 *M* KCl or liquid mercury to create an electrical connection between it and the chlorided silver wire in the conventional patch clamp electrode holder (*see* **Note 7**).
2. Connect the CFE to the electrode holder using an adaptor (*see* **Note 8**).
3. Immerse the CFE into the external saline solution for recording (*see* **Note 9**).
4. Place the electrode near a cell to be measured using a precision manipulator under a microscope with high magnification (×200–400). Because the shape of amperometric current is critically determined by the distance between the release and detection sites, a gentle touch of the cell membrane with the electrode produces the least-distorted signal (*see* **Notes 10** and **11**).
5. Measure the exocytotic events in a control condition and then with different stimuli.

3.2.1. Some Technical Comments for Amperometric Experiments

1. Holding potential for oxidation: potentials higher than the oxidation potential for the substrate should be applied to the electrode for efficient detection of catecholamines released from the cells. Typical electrode potentials are 400 mV for dopamine and 600 mV for serotonin. Most patch clamp amplifiers allow a holding potential of up to 200 mV. However, the potential can go higher by additional voltage supplied through an external voltage input.
2. Filtering and sampling: amperometric signals should be filtered, either digitally or with an analog filter before they are stored. A filter frequency up to 1–3 kHz is necessary to resolve fine details, such as rise time to the peak. Small signals (>1 pA) are detectable only after filtering at a lower frequency, such as 100 Hz, to remove the background high-frequency noise. However, strong filtering deforms the shape of fast amperometric signals. Sampling frequency should be at least twice and, better, four to five times the filter frequency based on the Nyquist criterion.
3. Change of solution: multiple solutions containing several test agents may be applied to the cell measured. If the solution perfusion pipet is close to the cell, then one needs to avoid abrupt change of flow that could generate damaging stress.
4. Solution level: gradual change of contact between the bath solution and the recording CFE causes a change of amperometric current level. Therefore, it is important to keep the bath solution level (i.e., vs the pipet level) constant. We control the level by removing the bath solution using a fine (25-gage) needle and high suction pressure. This continuously and rapidly removes the excessive perfused bath solution and keeps the solution level relatively constant.
5. Simultaneous measurement of multiple cellular signals: several parameters can be measured from a single cell for a better correlation between them. For example, cytosolic Ca^{2+} concentration can be monitored using Ca^{2+} sensitive dyes such as

indo-1 or fura-2 while exocytotic events are captured by amperometry from the same cell (**Fig. 1** and **refs.** *12* and *14*). Those experiments require multiple loading processes, and the sequence of loadings is important; in the case of loading by diffusion (**Subheading 3.3.2.**), cells are first loaded with Ca^{2+} dye and then the catecholamine. Otherwise, the oxidizable molecules can leak out during the loading of Ca^{2+} dye. For the loading by transporters (**Subheading 3.3.1.**), cells are loaded with the catecholamine in culture medium for several hours and sequentially with Ca^{2+} dye in Na^+-rich external solution for 20–40 min (*see* **Note 12**). In other cases, capacitance to measure exocytosis and Ca^{2+} current may be recorded along with amperometric events *(18)*.

3.3. Loading of Exogenous Oxidizable Molecules Into Secretory Vesicles

Amperometric measurements are normally limited to cells that package and secrete an endogenous oxidizable molecule such as dopamine or serotonin; however, in some cases oxidizable molecules can be introduced artificially *(19)*. For example, prolonged incubation of pancreatic β cells with 0.5–1 m*M* serotonin loads insulin granules with the neurotransmitter. The treatment promotes the accumulation of serotonin in insulin granules by specific cellular transporters. In addition, we have found that soaking a variety of cells in 50–100X higher concentrations of dopamine or serotonin forces the exogenous molecules to distribute into cytoplasm and acidic secretory vesicles by passive diffusion *(12,13,20)*. This loading process does not require specific transporters in the plasma membrane or vesicular membrane and therefore can be used, in principle, for most cell types. The exogenous molecules probably end up in many different intracellular acidic compartments, as suggested by Kim et al. *(13)*, addressing the mechanisms involved in this type of loading (*see* **Notes 13** and **14**).

3.3.1. Loading by Specific Transporters

1. Prepare 100 m*M* stock solution of the appropriate type of neurotransmitter that can be transported into the cytoplasm and secretory vesicles by endogenous cellular transporters (*see* **Note 15**).
2. Add 15 μL each neurotransmitter and ascorbic acid stock solutions to 1.5 mL culture medium (about 1 m*M* final concentration of both agents) and incubate cells for 4–10 h in the case of serotonin loading into insulin-secreting β cells *(19,21)* (*see* **Note 16**).
3. Perform amperometric experiments in an amine-free saline solution.

3.3.2. Loading by Diffusion

1. Incubate cells in 1–2 mL loading solution at room temperature. It is necessary to incubate cells for longer than 30 min in a loading medium containing 30–70 m*M* oxidizable catecholamines to detect significant oxidation currents *(13)*.

2. Transfer cells to the culture medium and keep at 37°C and 5% CO_2 for up to 4 h. Some cells, such as pancreatic duct epithelial cells, are kept in an amine-free Na^+-rich saline solution up to 2 h at room temperature before use.
3. Perform amperometric experiments in an amine-free saline solution (*see* **Note 17**).

3.3.3. Loading by Endocytosis

Secretory vesicles recycle into the cell by endocytosis. Incubation of cells with high concentrations of oxidizable molecules during and after the induction of evoked exocytosis can load recycling vesicle pools, and those "filled" vesicles then produce amperometric signal during the next exocytosis. We tested this protocol (fluid phase loading) with cerebellar granule cells *(13)* (*see* **Note 18**).

1. Stimulate the neurons with a loading solution containing 70 mM KCl plus 70 mM dopamine for 2 min.
2. Incubate cells in a dopamine-free, Na^+-rich saline solution for at least 30 s.
3. Measure exocytosis after activating a cell with K^+-rich (70 mM) solution.

3.4. Data Analysis

Amperometric results are analyzed in at least two different ways: integrating the currents to measure the total amount of secreted oxidizable molecules *(22)* or counting spikes to calculate the rate of exocytosis *(12,14)*. The former analysis is straightforward, so only the latter is discussed here. Counting spikes is possible only if spikes are so frequent they do not overlap each other.

1. Amperometric current recording is scanned by the algorithm described in **Subheading 2.**
2. After the automatic run, each detected event is inspected by eye and adjusted or removed if the judgment by the software is incorrect.
3. After detecting a spike occurrence, the algorithm saves time, area, and peak amplitude of the spike for further analysis.
4. To estimate the rate of exocytosis, the numbers of amperometric spikes per a selected time bin are counted, and a histogram is built (**Fig. 1B**). The time window should be decided depending on the kinetics of exocytosis of the cell under investigation.
5. To adjust for cell-to-cell variation and to describe the average cell behavior, the rate of exocytosis of each experiment is normalized to the baseline value before averaging (normalized rate of exocytosis).
6. To quantify the effect of the test agent on exocytosis, relative exocytosis is defined by the ratio of exocytosis rates in test and in control conditions.
7. Analysis of single events: several parameters, including peak amplitude and rise time to peak, can be estimated. In addition, the time integral of the current in single events yields a charge per event. Number of oxidizable molecules released from a vesicle can be calculated by dividing the charge by the elementary charge

of the electron (1.6×10^{-19} C) and the number of electrons emitted by the oxidation per molecule, such as two for dopamine (**Fig. 1A**).

4. Notes

1. Data can be saved in other types of storage devices, such as digital audio tape, and later replayed for digitization. Unless analysis of the original data at other filter frequencies is necessary, storage of data directly on hard disks is most convenient because of their current large capacity.
2. Ethanol removes the attached hydrophobic impurities and makes the insertion of the carbon fiber into the micropipet tip easier than in air. The fibers can be stored in 100% ethanol for longer than 6 mo without losing their catalytic properties.
3. To protect the brittle carbon fibers from damage, the bottom of the 60-mm plastic dish is covered by silicone elastomer, and the tips of the handling forceps are covered by short (~2-mm) sections of PE tubing.
4. It is advantageous to fill multiple (more than several tens) tips before going to the next step. The micropipet tip filled with a carbon fiber can be preserved for at least 6 mo, which is the longest time we have tested.
5. To yield better fixation of the rubber bulb, the inner surface of the forceps is scratched with coarse-grain sandpaper (50 grit).
6. A fabricated electrode can be stored at room temperature for a few months without noticeable change of sensitivity. A micropipet tip box can be used for the storage of multiple CFEs. Care must be taken so that the tip is not damaged during take-in and removal.
7. We prefer the KCl solution to mercury for safety reasons. The KCl solution is injected into the lumen of the micropipet tip using a fine plastic tip (Microloader, Eppendorf, Hamburg, Germany) to minimize damage to the carbon fiber inside. The silver wire is immersed approx 3 mm into the KCl solution in the CFE.
8. We cut the barrel of a 2-mL plastic syringe (BD Inc.), and the front part serves as the adapter. Inner diameter of the syringe fits tightly into the patch clamp electrode holders. The Luer fitting is ground so that it fits into the wide opening of the CFE.
9. When a holding potential of 600 mV is applied to the CFE, a large background current (>100 pA in the case of the 11-μm electrode) flows for an unknown reason, presumably because of oxidation of chemical groups. This ensures that the electrical connection between the CFE and the amplifier is correct. However, too large a leak current indicates excessive carbon fiber exposure, improper cutting, or damage to the plastic insulation. The current fades slowly within 20 min to reach a steady state of less than 10 pA. For long recordings (>10 min), it is necessary to let the background current fall as much as possible. Otherwise, a slow drift of the baseline current will appear when plotted as a compressed record. The "primed" electrodes can be used a few days later without losing sensitivity, and they reach the low leak current level much faster when immersed in the same saline solution again. Sensitivity and reaction time will fall with multiple uses of an electrode. The oxidation current induced by catecholamines critically depends

on the exposed area of the carbon fiber, ranging from 100 to 200 pA on the application of 10 μM norepinephrine to a new 11-μm CFE.

10. Avoid pushing too firmly to prevent damaging the cell membrane by the CFE. In epithelial cells, membrane stretch alone activates stretch-activated nonselective cation channels and increases both cytosolic Ca^{2+} concentration and Ca^{2+}-dependent exocytosis (S. R. Jung, unpublished data). It is critical to stop advancing the electrode whenever a shift of the cell membrane is observed. Therefore, a high optical resolution microscope is necessary for the optimal touch. We start the amperometric recording about 3 min after the touch to minimize any possible consequence of mechanical disturbance to the cell.

11. Flat cells growing on a thin substrate layer are especially vulnerable to damage by CFE. To reduce the mechanical stress, we use a thick collagen layer (Vitrogen, Cohesion Co., Palo Alto, CA) for pancreatic duct epithelial cells *(12,14)*. To improve the attachment of the material to a glass surface, a drop of Vitrogen:culture medium (1:1) mixture on a glass chip is dried 10 cm below the ultraviolet (UV) lamp in a laminar flow hood. UV light is supposed to form crosslinks between collagen fibers, but the assumption has not been tested experimentally. The Vitrogen layer forms a soft bed for cells when hydrated with the culture medium, and it stays on the chip for a few days.

12. Illuminating CFEs with UV light induces subpicoampere current artifacts that must be digitally corrected when larger than the baseline noise. Therefore, it is best to open the shutter that controls the UV excitation of Ca^{2+} dyes only briefly.

13. In choosing the neurotransmitter to be loaded, a few factors need to be considered. First, avoid a neurotransmitter for which the cell has a membrane receptor. Secretion of the neurotransmitter can activate the intracellular signals in an autocrine fashion. Second, some cells have adverse reactions to a certain type of neurotransmitter; for example, cultured pinealocytes became swollen to some extent after the loading of 70 mM serotonin but not dopamine. The effect may be mediated by a serotonin receptor expressed in the cells *(23)*.

14. Exocytosis of secretory vesicles after the exogenous loading in nonexcitable cells gives amperometric events similar to those in neurons and endocrine cells, including the "foot" events and leak of transmitters through fusion pores. Hence, the molecular events for exocytosis in diverse cell types are similar *(12,15)*. The nonspecific diffusion method could load all tested cells, including AtT-20, PC-12, pituitary gonadotrope, cerebellar granule and dorsal root ganglion neurons, pancreatic epithelial cells, and yeast *(12,13,24)*.

15. Low concentration of precursors for oxidizable neurotransmitters can be used if a cell is equipped with an appropriate converting enzyme and a vesicular transporter. For example, a PC-12 cell incubated with 50 μM L-DOPA, precursor of dopamine, for 40–70 min displayed increased amperometric signals *(25)*.

16. The incubation time varies depending on the transport rate of the molecules. The best incubation time for a specific cell must be determined by measuring exocytosis evoked by an appropriate stimulus, for example, K^+-rich solution to excitable cells expressing voltage-gated Ca^{2+} channels.

17. Amperometric signals typically decrease over 4–5 h after loading is stopped, presumably because of a loss of vesicular amines through spontaneous exocytosis and by leak of cytoplasmic amines to the culture medium.
18. This method can also be used for other types of cells for which stimuli other than high K^+ are used. For example, addition of 10 μM forskolin, a strong initiator of the duct epithelial exocytosis, to the loading solution (**Subheading 3.3.2.**) increases the peak size and charge of quantal events. If the incubation with the stimulants is short, then this loading protocol may mark the vesicles designated for evoked exocytosis; the nonspecific loading method (**Subheading 3.3.2.**) marks all kinds of secretory vesicles.

Acknowledgments

I would like to thank L. Miller, J. G. Duman, and B. Hille for the comments on the manuscript and S.R. Jung for providing the recording shown in **Fig. 1**. This work was supported by the R&D Program of Advanced Technologies from the Ministry of Commerce, Industry, and Energy of Korea and a National Institutes of Health grant (AR17803).

References

1. Hille, B. (2001) *Ion Channels of Excitable Membranes,* Sinauer, Sunderland, MA.
2. Berridge, M. J., Bootman, M. D., and Roderick, H. L. (2003) Calcium signalling: dynamics, homeostasis and remodelling. *Nat. Rev. Mol. Cell. Biol.* **4,** 517–529.
3. Gillis, K. D. (1995) Techniques for membrane capacitance measurements, in *Single Channel Recording* (Sakmann, B., and Neher, E., eds.), Plenum, New York, pp. 155–198.
4. Cochilla, A. J., Angleson, J. K., and Betz, W. J. (1999) Monitoring secretory membrane with FM1-43 fluorescence. *Annu. Rev. Neurosci.* **22,** 1–10.
5. Wightman, R. M., Jankowski, J. A., Kennedy, R. T., et al. (1991) Temporally resolved catecholamine spikes correspond to single vesicle release from individual chromaffin cells. *Proc. Natl. Acad. Sci. USA* **88,** 10,754–10,758.
6. Chow, R. H. and von Rüden, L. (1995) Electrochemical detection of secretion from single cells, in *Single Channel Recording* (Sakmann, B., and Neher, E., eds.), Plenum, New York, pp. 245–275.
7. Kawagoe, K. T., Zimmerman, J. B., and Wightman, R. M. (1993) Principles of voltammetry and microelectrode surface states. *J. Neurosci. Methods* **48,** 225–240.
8. Schroeder, T. J., Jankowski, J. A., Senyshyn, J., Holz, R. W., and Wightman, R. M. (1994) Zones of exocytotic release on bovine adrenal medullary cells in culture. *J. Biol. Chem.* **269,** 17,215–17,220.
9. Bruns, D. and Jahn, R. (1995) Real-time measurement of transmitter release from single synaptic vesicles. *Nature* **377,** 62–65.
10. Jaffe, E. H., Marty, A., Schulte, A., and Chow, R. H. (1998) Extrasynaptic vesicular transmitter release from the somata of substantia nigra neurons in rat midbrain slices. *J. Neurosci.* **18,** 3548–3553.

11. Zhou, Z. and Misler, S. (1995) Action potential-induced quantal secretion of catecholamines from rat adrenal chromaffin cells. *J. Biol. Chem.* **270,** 3498–3505.
12. Koh, D. S., Moody, M. W., Nguyen, T. D., and Hille, B. (2000) Regulation of exocytosis by protein kinases and Ca^{2+} in pancreatic duct epithelial cells. *J. Gen. Physiol.* **116,** 507–520.
13. Kim, K. T., Koh, D. S., and Hille, B. (2000) Loading of oxidizable transmitters into secretory vesicles permits carbon-fiber amperometry. *J. Neurosci.* **20,** RC101.
14. Jung, S. R., Kim, M. H., Hille, B., Nguyen, T. D., and Koh, D. S. (2004) Regulation of exocytosis by purinergic receptors in pancreatic duct epithelial cells. *Am. J. Physiol. Cell. Physiol.* **286,** C573–C579.
15. Zhou, Z., Misler, S., and Chow, R. H. (1996) Rapid fluctuations in transmitter release from single vesicles in bovine adrenal chromaffin cells. *Biophys. J.* **70,** 1543–1552.
16. Schulte, A. and Chow, R. H. (1996) A simple method for insulating carbon-fiber microelectrodes using anodic electrophoretic deposition of paint. *Anal. Chem.* **68,** 3054–3058.
17. Koh, D. S. and Hille, B. (1999) Rapid fabrication of plastic-insulated carbon-fiber electrodes for micro-amperometry. *J. Neurosci. Methods* **88,** 83–91.
18. Moser, T., Chow, R. H., and Neher, E. (1995) Swelling-induced catecholamine secretion recorded from single chromaffin cells. *Pflügers Arch.* **431,** 196–203.
19. Smith, P. A., Duchen, M. R., and Ashcroft, F. M. (1995) A fluorimetric and amperometric study of calcium and secretion in isolated mouse pancreatic beta-cells. *Pflügers Arch.* **430,** 808–818.
20. Billiard, J., Koh, D. S., Babcock, D. F., and Hille, B. (1997) Protein kinase C as a signal for exocytosis. *Proc. Natl. Acad. Sci. USA* **94,** 12,192–12,197.
21. Chen, L., Koh, D. S., and Hille, B. (2003) Dynamics of calcium clearance in mouse pancreatic β-cells. *Diabetes* **52,** 1723–1731.
22. Koh, D. S. and Hille, B. (1997) Modulation by neurotransmitters of catecholamine secretion from sympathetic ganglion neurons detected by amperometry. *Proc. Natl. Acad. Sci. USA* **94,** 1506–1511.
23. Miguez, J. M., Simonneaux, V., and Pevet, P. (1997) The role of the intracellular and extracellular serotonin in the regulation of melatonin production in rat pinealocytes. *J. Pineal Res.* **23,** 63–71.
24. Zhang, C. and Zhou, Z. (2002) Ca^{2+}-independent but voltage-dependent secretion in mammalian dorsal root ganglion neurons. *Nat. Neurosci.* **5,** 425–430.
25. Pothos, E., Desmond, M., and Sulzer, D. (1996) L-3,4-dihydroxyphenylalanine increases the quantal size of exocytotic dopamine release in vitro. *J. Neurochem.* **66,** 629–636.

IV

METHODS FOR STUDYING CHANNELOPATHIES, GENETIC SCREENING, AND MOLECULAR BIOLOGY

13

Genetic Screening for Functionality of Bacterial Potassium Channel Mutants Using K⁺ Uptake-Deficient *Escherichia coli*

Lyubov V. Parfenova and Brad S. Rothberg

Summary

Potassium channels play an essential role in a wide range of biological processes, including cell volume regulation and the maintenance and control of electrical signals. With the advent of the structural era of ion channel biology, it has become critical to learn more about the functional properties of the prokaryotic channels, and this is the area in which genetic screens have become an increasingly useful approach. Here, we describe a bacteria-based complementation assay that we applied to investigate gating mutants of the prokaryotic K⁺ channel MthK, which was cloned from the archeon *Methanobacterium thermoautotrophicum*. The results demonstrated that heterologously expressed MthK is fully assembled and functional in *Escherichia coli*. This complementation assay should be useful in the initial identification of prokaryotic K⁺ channel mutants that result in altered channel function.

Key Words: Complementation; genetic screen; heterologous expression; MthK.

1. Introduction

Potassium (K⁺) channels are ubiquitous membrane proteins that perform a variety of specific functions in different organisms, including the maintenance of the resting membrane potential and control of electrical excitability. Our knowledge of K⁺ channel structure has exploded since 1998, in large part because of the cloning and high-level expression of K⁺ channels from prokaryotes such as *Streptomyces lividans*. It is this ability to overexpress prokaryotic channels in prokaryotic expression systems that has led to crystallographic studies and high-resolution structural data (*1*). These exciting new structures often beautifully dovetail with current working hypotheses of channel function.

From: *Methods in Molecular Biology, vol. 337: Ion Channels: Methods and Protocols*
Edited by: J. D. Stockand and M. S. Shapiro © Humana Press Inc., Totowa, NJ

With the advent of the structural era of ion channel biology, it has become critical to learn more about the functional properties of the bacterial channels, and it is in this area that genetic screens have become an increasingly useful tool. A single genetic screen can enable the rapid isolation of several potentially interesting mutations that produce altered channel behavior. The interesting mutant channels revealed by an initial screen can then be characterized in more detail using biochemical and electrophysiological methods. Screens using complementation of K+ uptake-deficient yeast have been useful in studying the Kat1 and AKT1 channels from *Arabidopsis*, as well as mammalian G protein-activated K+ channels *(2,3)*. In addition, several investigators have used K+ uptake-deficient strains of *Escherichia coli* to study gating in prokaryotic K+ channels such as KcsA and MjK *(4,5)*.

Here, we describe a genetic screen that we are using to investigate potential gating mutants of the prokaryotic K+ channel MthK, which was cloned from the archeon *Methanobacterium thermoautotrophicum (6)*. As with previously described bacterial K+ channel screens, we exploit strains of *E. coli* in which the three high-affinity K+-uptake systems (Kdp, Kup, and Trk) have been eliminated *(7,8)*. Normally, these strains cannot survive in low K+ media such as Luria-Bertani broth. However, they can be rescued either by elevating [K+] in the growth medium or through heterologous expression of a functional (i.e., "open") K+ channel in the strain *(4,5)*.

We expressed MthK in the TK2446 strain (kindly provided by Wolf Epstein, University of Chicago, Illinois) and found that it results in complementation (**Fig. 1**); thus, future screens could be performed to reveal mutations in MthK that reduce channel conductance or open probability. As negative and positive controls, we used KcsA and the KcsA mutant A108S, respectively. It was shown previously that KcsA, which rarely opens except at very low pH, yields no complementation in a related K+ uptake-deficient *E. coli* strain; the A108S mutant, which opens much more than wild type, yields robust complementation *(4)*.

We found that MthK channel expression and complementation in TK2446 was independent of the type of promoter present in the expression vector. Also, expression of the MthK channel's C-terminal cytoplasmic domain did not result in complementation, demonstrating that the full-length channel is required to allow K+ conduction in the TK2446 strain. MthK expression and oligomeric assembly in the bacteria were confirmed by Western blot analysis performed using crude bacterial lysates (**Fig. 2**).

2. Materials

2.1. Plasmids and E. coli *Strains*

1. *E. coli* strains TK 2446: F- *thi rha lacZ nagA* Δ*(kdp FAB)5 trkD1 trkG(kan) trkH(cam)* Δ*(trkA-mscL')*, kindly provided by Wolf Epstein (University of Chicago).

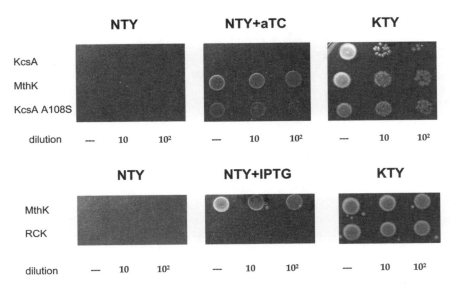

Fig. 1. (**A**) MthK expression complements growth in low K⁺ (NTY) in TK2446 cells. KcsA and the KcsA mutant A108S were used as negative and positive controls, respectively. Channel expression was induced with anhydrotetracycline (aTC). All transformants survived in high K⁺ (KTY). (**B**) MthK expression complements growth in low K⁺ in TK2446 cells; the MthK cytoplasmic RCK domain (expressed without the pore-forming transmembrane segments) does not. Channel expression was induced with IPTG. All transformants survived in high K⁺ (KTY).

2. MthK cDNA, kindly provided by Christopher Miller (HHMI, Brandeis University Waltham, MA), was subcloned into the pASK-90 and modified pQE-82L vectors (*see* **Note 1**).
3. KcsA in pASK-90, kindly provided by Christopher Miller, was used as a negative control.
4. KcsA mutant A108S was generated using QuickChange (Stratagene, La Jolla, CA) and used as a positive control.

2.2. Preparation and Transformation of E. coli-Competent Cells

1. High-K⁺ liquid medium (KLM): autoclave 10 g tryptone, 5 g yeast extract, and 10 g KCl per 1 L H₂O.
2. KLM plates: autoclave 10 g tryptone, 5 g yeast extract, 10 g KCl, and 10 g agar per 1 L H₂O. Allow the medium to cool to approx 50°C, add ampicillin, mix by swirling, and pour into the plates.
3. Ampicillin (1000X solution): 1 g per 10 mL H₂O, sterilize through 0.22-μm filter, and store in 1-mL aliquots at –20°C.
4. Transformation buffer (TB): 10 m*M* PIPES, 55 m*M* MnCl₂, 15 m*M* CaCl₂, and 250 m*M* KCl (*see* **Note 2**).

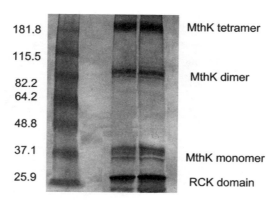

181.8	MthK tetramer
115.5	
82.2	MthK dimer
64.2	
48.8	
37.1	MthK monomer
25.9	RCK domain

Fig. 2. Western blot analysis of MthK wild-type expression in TK2446 cells. The blot displays bands corresponding to the RCK domain (26 kDa) and monomer (~38 kDa), dimer (~90 kDa), and tetramer/octamer (200 kDa). **Lane 1** contains crude lysate from preinduced cells (KLM medium). **Lanes 2** and **3** contain crude lysate from cells harvested postinduction in KLM or LKLM medium, respectively.

2.3. E. coli *Complementation Assay*

1. Low-K⁺ liquid medium (LKLM): autoclave 10 g tryptone, 5 g yeast extract, 10 g NaCl per 1 L H_2O, then add 400 µL 2.5 M KCl (sterile filtered).
2. KLM medium.
3. TY-agar: 10 g tryptone, 5 g yeast extract, and 10 g agar per 1 L H_2O; autoclave.
4. KTY plates: for one plate, put 800 µL 2.5 M KCl and 60 µL 5 M NaCl into a sterile 50-mL conical tube. Add TY-agar (cooled to ~50°C) to 20 mL final volume. Add 40 µL ampicillin, mix briefly, and pour into the Petri dish immediately.
5. NTY plates: for one plate, put 460 µL 5 M NaCl into a 50-mL conical tube. Add TY-agar to 20 mL final volume. Add 40 µL ampicillin, mix briefly, and pour into the Petri dish immediately (*see* **Note 3**).
6. Isopropyl-β-D-1-thiogalactoside (IPTG) (100X solution): prepare 100 mM IPTG solution in water, sterilize through 0.22-µm filter, aliquot, and store at −20°C.
7. Anhydrotetracycline (aTC) (1000X solution): dissolve 20 mg aTC in 100 mL dimethylformamide, store at −20°C in an amber glass bottle.

2.4. Sodium Dodecyl Sulfate Polyacrylamide Gel Electrophoresis and Western Blotting for MthK

1. Protein lysis/denaturing buffer: 20 mM Tris-HCl pH 6.0, 4 M urea, and 2% (w/v) sodium dodecyl sulfate (SDS).
2. Running buffer: 25 mM Tris-HCl, 192 M glycine, and 0.1% (w/v) SDS, pH 8.3. Store at room temperature.
3. Precast 10% polyacrylamide electrophoresis gel from Gradipore (Frenchs Forest NSW, Australia).

4. Laemmli sample buffer: 75 mM Tris-HCl pH 6.8, 2% (w/v) SDS, 7.5% (w/v) glycerol, 200 mM β-mercaptoethanol, and 0.03% (w/v) bromophenol blue.
5. Prestained molecular weight markers: BenchMark prestained protein ladder from Invitrogen (Carlsbad, CA).
6. Transfer buffer: 25 mM Tris and 0.192 M glycine, pH 8.3. Store at room temperature.
7. Pure nitrocellulose membrane (0.45 μm) (Bio-Rad, Hercules, CA).
8. Minitransblot filter paper from (Bio-Rad).
9. Tris-buffered saline (TBS): 0.5 M NaCl and 20 mM Tris-HCl pH 7.5.
10. Tris-buffered saline with Tween (TBST): 0.5 M NaCl, 20 mM Tris-HCl pH 7.5, and 0.1% (w/v) Tween-20.
11. Blocking/incubation buffer: 5% (w/v) Carnation nonfat dry milk (Nestle USA, Solon, OH) in TBST.
12. Primary antibody: Penta-His™ antibody (mouse monoclonal immunoglobulin G$_1$) from Qiagen (Valencia, CA); dissolve in TBS to final concentration of 0.1 mg/mL.
13. Secondary antibody: liquid affinity purified peroxidase-labeled goat antimouse conjugate (1 mg/mL) from Kirkegaard and Perry (Gaithersburg, MD).
14. Peroxidase 3,3'-diaminobenzidine tetrahydrochloride (DAB) substrate kit from Vector Laboratories (Burlingame, CA).

3. Methods

Although screens of other prokaryotic K+ channels have used the TK2420 and LB2003 K+ uptake-deficient strains *(4,5,7,8)*, in our studies with MthK it was critical to use the TK2446 strain. We found in our initial screens that complementation of the TK2420 strain can be obtained with a secondary product of the MthK gene that consists only of a cytoplasmic domain of the channel subunit (the RCK domain) without the pore-8209;forming transmembrane portion *(6)*. It seems that, in the case of MthK, the similarity of this RCK domain with the native *E. coli* TrkA protein enables it to substitute for TrkA (which is deleted in the TK2420 strain) and, along with the TrkG and TrkH proteins, reconstitute the Trk K+ uptake system, resulting in complementation. Complementation by the MthK RCK domain is eliminated in TK2446, which also contains deletions of both TrkG and TrkH.

3.1. Preparation of Competent E. coli Cells

1. Inoculate 3 mL KLM medium with TK2446 *E. coli* cells and incubate the culture overnight at 37°C.
2. Add 2.5 mL of the overnight culture to 250 mL KLM medium and incubate the culture at 25–30°C to mid-log phase (until the absorbance at 600 nm is approx 0.4–0.6).
3. Chill the culture on ice (at least 10 min).
4. Spin at 2500g (5000 rpm in a Sorvall GSA or 3000 rpm in a Beckman J-6B centrifuge) for 10 min at 4°C.

5. Resuspend cells by gentle swirling in 80 mL ice-cold TB.
6. Incubate the cell suspension on ice for 20 min (20–60 min).
7. Spin at 2500g for 10 min at 4°C.
8. Resuspend the pellet gently in 18.8 mL ice-cold TB and add 1.4 mL dimethyl sulfoxide by drops, swirling the cell suspension gently.
9. Incubate the cell suspension on wet ice for at least 10 min.
10. Aliquot the cell suspension at 500 μL per sterile microcentrifuge tube.
11. Shock-freeze the cell suspension in liquid nitrogen and store the tubes at 80°C or in liquid nitrogen.

3.2. E. coli *Transformation*

1. Thaw competent cells on ice.
2. Add 1 μL of 1 μg/μL plasmid DNA solution to 100 μL competent cell suspension; mix gently. Use 100 μL of cell suspension with 1 μL water as a negative control.
3. Incubate the tubes on ice for 20 min.
4. Heat shock the cells for 1 min at 42°C.
5. Place the tubes immediately on ice for 2 min.
6. Add 900 μL KLM medium to each tube.
7. Incubate for 1 h at 37°C with vigorous shaking (on a shaker table at 225–250 rpm).
8. Centrifuge cells at 5000g for 5 min; remove all but approx 50 μL supernatant.
9. Resuspend cell pellet with the supernatant remaining in the tube by pipetting.
10. Plate the suspension on KLM agar plates containing antibiotic (ampicillin).
11. Incubate the plates overnight at 37°C.

3.3. E. coli *Complementation Assay*

1. Inoculate transformed TK2446 *E. coli* cells in KLM/ampicillin; grow overnight at 37°C with vigorous shaking.
2. Put 100 μL of overnight culture in 5 mL KLM/ampicillin in a culture tube; grow at 37°C with vigorous shaking to mid-log phase.
3. Take a 1-mL sample of cell suspension immediately before induction, centrifuge the sample (12,000g, 2 min), and freeze the pellet for Western blot analysis.
4. To induce heterologous protein expression, add 40 μL IPTG or 4 μL aTC and incubate at 37°C for 3 h.
5. At the end of the induction period, take a 1-mL sample of cell suspension, centrifuge the sample (12,000g, 2 min), and freeze the pellet for Western blot analysis.
6. Check the optical density at 600 nm (OD_{600}) of the culture after induction and normalize the density to 0.5 with sterile water. This is the undiluted cell suspension.
7. Prepare two dilution tubes for each sample: put 900 μL sterile water into each tube, add 100 μL from the undiluted cell suspension into the first tube (10^1 dilution), mix briefly, and transfer 100 μL of this new suspension to the other tube (10^2 dilution).

8. Spot 3.5-µL drops of cell suspension from the undiluted cell suspension and the dilution tubes on NTY and KTY plates with and without inductor (aTC or IPTG) (*see* **Note 4**).
9. Incubate plates overnight at 37°C until colonies form.

3.4. Growing Cells for Checking MthK Expression in LKLM Medium

1. Take 100 µL normalized cell culture from **Subheading 3.3.**, **step 6** to a sterile Eppendorf tube.
2. Pellet cells at 12,000g for 2 min and remove the supernatant by pipet. The pellet should be very small. Take care to throw out the supernatant and try not to disturb the pellet with the pipet tip.
3. Wash the pellet by LKLM: resuspend the pellet in 0.5 mL LKLM, spin down, and remove the supernatant carefully. This is a critical step to remove the traces of KLM, which contain high K⁺ concentration.
4. Resuspend the pellet in 0.5 mL LKLM and put the suspension in 4.5 mL LKLM/ ampicillin tube with inductor (aTC or IPTG).
5. Grow at 37°C, shaking vigorously until the OD$_{600}$ reaches 1.0.
6. Take a 1-mL sample, spin down the cells (12,000g, 2 min), discard the supernatant, and save the pellet for Western blot analysis.

3.5. SDS Polyacrylamide Gel Electrophoresis and Western Blotting of MthK

1. To prepare the crude protein extract samples from *E. coli* cells, thaw pellets from **Subheading 3.3.**, **steps 3–6**.
2. Add 200 µL lysis/denaturing buffer; vortex to make a homogeneous suspension. Incubate at room temperature 10 min.
3. Centrifuge the lysates at 10,000g for 20 min to pellet cellular debris. Transfer the supernatants to new tubes.
4. Add 10 µL 2X SDS polyacrylamide gel electrophoresis sample buffer to 10 µL supernatant, mix by pipetting, and load into the gel wells immediately. *Do not boil or heat samples* (*see* **Note 5**). Reserve one well for prestained molecular markers.
5. Start the gel running at 50 V. When samples enter the gel completely, increase the voltage to 150 V.
6. Run the gel until the 24-kDa marker reaches the bottom of the gel.
7. Remove the gel and wash with transfer buffer.
8. Prepare the transfer cassette according to the manufacturer's instructions.
9. Run at 350 mA for 60 min.
10. Once the transfer is complete, take the cassette out of the tank and carefully disassemble. Remove the nitrocellulose and make sure that the colored molecular weight markers are visible on the membrane.
11. Wash the membrane with TBS and put it in 20 mL blocking/incubation buffer.
12. Incubate for 60 min on a rocker platform; then remove all but 5 mL of the buffer.
13. Add 2.5 µL primary antibody, swirl gently, and incubate at +4°C overnight (in the refrigerator).

14. In the morning, incubate the membrane for 30 min at room temperature on a rocker platform.
15. Remove the blocking buffer and wash the membrane three times for 5 min each with 15 mL TBST.
16. Dissolve 1 μL secondary antibody in 20 mL blocking/incubation buffer; pour into a container and incubate 60 min.
17. Remove the secondary antibody solution from the container.
18. Wash the membrane three times for 5 min each with 15 mL TBS.
19. Mix DAB kit reagents according the manufacturer's instructions.
20. Add the DAB solution to the container and swirl until the colored protein bands appear.

4. Notes

1. Plasmid cDNA is expressed from a prokaryotic promoter.
2. Adjust to pH 6.7 with 5 N KOH prior to adding $MnCl_2$ to avoid formation of an insoluble salt complex.
3. Elemental analysis of agar used indicates the presence of 1 mM K$^+$, so it is not necessary to add KCl to a final 1 mM K$^+$ concentration into the NTY plates *(2)*.
4. Add inductor (aTC or IPTG) during preparation of NTY plates (**Subheading 2.3.**, **step 5**). We advise adding aTC or IPTG solutions after filling the 50-mL tube with cooled NT.
5. MthK full-length tetramer is stable in the presence of detergents (even SDS), but boiling prior to SDS polyacrylamide gel electrophoresis results in aggregation.

Acknowledgments

We wish to thank Lise Heginbotham (Yale University) for her gift of the TK2420 strain, which was used in preliminary experiments; Crina Nimigean (University of CA-Davis) for providing constructs and much advice on working with MthK and KcsA; and Brittany Crane for technical assistance. This work was supported by R01 GM68523 to B. S. R.

References

1. Doyle, D. A., Morais Cabral, J., Pfuetzner, R. A., et al. (1998) The structure of the potassium channel: molecular basis of K$^+$ conduction and selectivity. *Science* **280,** 69–77.
2. Bertl, A., Anderson, J. A., Slayman, C. L., Sentenac, H., and Gaber, R. F. (1994) Inward and outward rectifying potassium currents in *Saccharomyces cerevisiae* mediated by endogenous and heterologously expressed ion channels. *Folia Microbiol. (Praha)* **39,** 507–509.
3. Yi, B. A., Lin, Y. F., Jan, Y. N., and Jan, L. Y. (2001) Yeast screen for constitutively active mutant G protein-activated potassium channels. *Neuron* **29,** 657–667.
4. Irizarry, S. N, Kutluay, E., Drews, G., Hart, S. J., and Heginbotham, L. (2002) Opening the KcsA K$^+$ channel: tryptophan scanning and complementation analysis lead to mutants with altered gating. *Biochemistry* **41,** 13,653–13,662.

5. Hellmer, J., and Zeilinger, C. (2003) MjK1, a K⁺ channel from *M. jannaschii*, mediates K⁺ uptake and K⁺ sensitivity in *E. coli*. *FEBS Lett.* **547,** 165–169.

6. Jiang, Y., Lee, A., Chen, J., Cadene, M., Chait, B. T., and MacKinnon, R. (2002) Crystal structure and mechanism of a calcium-gated potassium channel. *Nature* **417,** 515–522.

7. Schlösser, A., Meldorf, M., Stumpe, S., Bakker, E. P., and Epstein, W. (1995) TrkH and its homolog, TrkG, determine the specificity and kinetics of cation transport by the Trk system of *Escherichia coli*. *J. Bacteriol.* **177,** 1908–1910.

8. Bossemeyer, D., Borchard, A, Dosch, D. C., et al. (1989) K⁺-transport protein TrkA of *Escherichia coli* is a peripheral membrane protein that requires other *trk* gene products for attachment to the cytoplasmic membrane. *J. Biol. Chem.* **264,** 16,403–16,410.

14

KCNQ1 K+ Channel-Mediated Cardiac Channelopathies

Gildas Loussouarn, Isabelle Baró, and Denis Escande

Summary

KCNQ1 is a voltage-activated potassium channel α-subunit expressed in various cell types, including cardiac myocytes and epithelial cells. KCNQ1 associates with different β-subunits of the KCNE protein family. In the human heart, KCNQ1 associates with KCNE1 to generate the I_{Ks} current characterized by its slow activation and deactivation kinetics. Mutations in either KCNQ1 or KCNE1 are responsible for at least four channelopathies that lead to cardiac dysfunction and one that leads to congenital deafness: the Romano-Ward syndrome, the short QT syndrome, atrial fibrillation, and the Jervell and Lange-Nielsen syndrome (cardioauditory syndrome). To date, nearly 100 different KCNQ1 mutations have been reported as responsible for the cardiac long QT syndrome, characterized by prolonged QT interval, syncopes, and sudden death. Patch clamp and immunofluorescence techniques are instrumental for characterization of the molecular mechanisms responsible for the altered function of KCNQ1 and its partners.

Key Words: Atrial fibrillation; COS-7 cells; immunofluorescence; KCNQ1; long QT syndrome; patch clamp; short QT syndrome.

1. Introduction

In the heart, KCNQ1, associated with KCEI, a β-subunit, generates the slow component of the delayed rectifier K+ current, I_{Ks} *(1,2)*. Growing evidence suggests that KCNQ1 is part of a channel complex *(3–5)*. We have previously reported an alternative splice cardiac isoform of KCNQ1 (isoform 2; *6*) that acts as an endogenous dominant negative partner on the channel isoform (isoform 1). Isoform 2 mRNA may represent up to 32% of the total amount of KCNQ1 mRNA in human midmyocardial tissue *(7)*. We proposed that expression of isoform 2 in the midmyocardium is responsible for the decrease in I_{Ks} and the prolonged action potential.

From: *Methods in Molecular Biology, vol. 337: Ion Channels: Methods and Protocols*
Edited by: J. D. Stockand and M. S. Shapiro © Humana Press Inc., Totowa, NJ

Mutation in any protein belonging to the channel complex (including KCNQ1) may alter the channel function and lead to channelopathy. Loss-of-function mutations in the KCNQ1 gene have been recognized as the most frequent cause for the autosomic dominant (Romano-Ward) form of the long QT syndrome, a life-threatening familial disorder characterized by prolonged cardiac repolarization *(8)*. KCNQ1 and KCNE1 were also demonstrated as responsible for the recessively inherited Jervell and Lange-Nielsen cardioauditory syndrome, characterized by bilateral deafness associated with prolonged cardiac repolarization *(9)*. Two other cardiac channelopathies linked to a gain-of-function mutation in KCNQ1 have been reported: short QT syndrome *(10)* and atrial fibrillation *(11)*. The molecular mechanisms potentially leading to the channelopathy are multiple and require various methods of investigation:

1. An alteration of the protein synthesis or degradation can be assayed by Western blotting and immunocytochemistry.
2. Altered channel trafficking related to a mutation in either KCNQ1 or one of its partners can be investigated by immunochemistry and patch clamp experiments: (1) the level of immunofluorescence is a rough estimate of protein targeting to the plasma membrane; (2) the current density measured in the whole-cell configuration of the patch clamp technique is a good indicator of membrane protein expression, provided that the channel characteristics ($V_{0.5}$, activation and deactivation kinetics) are not too profoundly altered by the mutation (*see* **Note 1**).
3. Altered channel function related to a mutation in KCNQ1 can be investigated in detail by various configurations of the patch clamp technique.
4. Altered channel function related to a mutation in a regulatory protein can be assessed with patch clamp by comparing the channel characteristics in the presence of the wild-type (WT) or mutant regulatory protein.

Immunocytochemistry and patch clamp techniques *(12,13)* have been described in detail many times. This chapter focuses on the specific adaptation of these techniques to study the molecular mechanisms underlying KCNQ1-mediated channelopathies.

2. Materials

2.1. Cell Culture

1. African green monkey kidney-derived cell line COS-7 (American Type Culture Collection, Rockville, MD). Unlike HEK293 cells *(14)*, COS-7 cells do no show endogenous K^+ currents, which may be confounding for studying recombinant I_{Ks} current.
2. Dulbecco's modified Eagle's medium supplemented with 10% fetal calf serum and antibiotics (100 IU/mL penicillin and 100 mg/mL streptomycin; all from Gibco, Paisley, UK).

3. Solution of trypsin (0.05%) and ethylenediaminetetraacetic acid (0.02%) (Sigma, St. Louis, MO).
4. Dulbecco's phosphate-buffered saline (10X), liquid calcium and magnesium free (Gibco).
5. Glass cover slips (Marienfeld, Lauda-Koenigshofen, Germany).
6. 35-mm Petri dish (Greiner Bio-One, Frickenhausen, Germany).

2.2. Plasmid Transfection and Microinjection

1. Human KCNQ1 (isoforms 1 and 2) and KCNE1 complementary deoxyribo-nucleic acids (DNAs), subcloned into the mammalian expression vectors pCI *(6,15)* and pCR *(6)*, respectively (Promega, Madison, WI) under the control of a cytomegalovirus enhancer/promoter.
2. pEGFP-N2-KCNQ1 constructed by subcloning the KCNQ1 complementary DNAs lacking the stop codon into the pEGFP-N2 plasmid (coding for the green fluorescent protein [GFP]; Clontech, Palo Alto, CA) and expressing the tagged protein KCNQ1-GFP. Used to localize the protein in nonpermeabilized cells, including intracellular compartments.
3. pCI-KCNQ1-HA expressing the channel isoform flagged with 2 HA epitopes in the extracellular loop between transmembrane domains S3 and S4 at positions K217 and Q220. These insertions modify the sequence between S3 and S4 to 216GSKYPYDVPDYAGQYPYDVPDYAVFA223, as verified by sequencing. This is used to detect the channel localized at the cytoplasmic membrane only.
4. pDsRed-ER (endoplasmic reticulum) vector, containing the KDEL sequence of the calreticulin protein cloned in frame with the DsRed sequence of pDsRed (a gift from D. J. Snyders *[16]*), is used to visualize the endoplasmic reticulum.
5. pEGFP-N3 plasmid used as a reporter gene (Clontech).
6. Plasmids are diluted in a buffer containing 50 mM HEPES, 50 mM NaOH, and 40 mM NaCl at pH 7.4.
7. JetPEI (Polyplus-Transfection, Illkirch, France) or Fugene-6 (Roche Molecular Biochemical, Basel, Switzerland) used as transfecting agents.
8. VectaSpin microcentrifuge tubes (0.45 µM polysulfone) (Whatmann, Kent, UK).
9. The microinjector 5246 system and the micromanipulator 5171 system (Eppendorf, Hamburg, Germany) are mounted on an Eclipse TE200 inverted micro-scope (Nikon, Tokyo, Japan). Injection pipets are pulled from borosilicate glass capillaries (Harvard Apparatus, Edenbridge, UK) with a P-87 puller (Sutter Instruments, Novato) and filled with a microloader (Eppendorf). Alternatively, ready-to-use injection pipets are available (Eppendorf).

2.3. Western Blot

1. Phosphate-buffered saline (PBS): 135 mM NaCl, 4.1 mM KCl, 0.8 mM NaH$_2$PO$_4$, and 5 mM Na$_2$HPO$_4$ pH 7.4.
2. Sodium dodecyl sulfate (SDS) gel-loading buffer: 50 mM Tris-HCl pH 6.8, 100 mM dithiothreitol, 2% SDS, 0.1% bromophenol blue, and 10% glycerol.
3. Tris-buffer, 25 mM, for electroblotting 40.7 mM glycine and 20% (v/v) methanol at pH 8.3.

2.4. Antibodies

1. For Western blot, mouse monoclonal anti-HA antibody (clone 12CA5, Boehringer Mannheim).
2. For Western blot, biotin-conjugated goat antimouse antibody (Sigma).
3. For immunocytochemistry, rabbit monoclonal anti-HA antibody (Sigma).
4. For immunocytochemistry, tetramethylrhodamine isothiocyanate-conjugated goat antirabbit immunoglobulin G (Sigma).

2.5. Patch Clamp

1. Standard extracellular medium for the whole-cell and permeabilized patch (also called perforated patch) configurations: 145 mM NaCl, 4 mM KCl, 1 mM MgCl$_2$, 1 mM CaCl$_2$, 5 mM HEPES, and 5 mM glucose; adjust to pH 7.4 with NaOH.
2. Microperfusion (extracellular) medium for the whole-cell and permeabilized patch configuration: 145 mM Na-gluconate, 4 mM K-gluconate, 7 mM Ca-gluconate (1 mM free Ca^{2+}), 4 mM Mg-gluconate (1 mM free Mg^{2+}), 5 mM HEPES, and 5 mM glucose; adjust to pH 7.4 with NaOH.
3. Pipet (intracellular) medium for the whole-cell patch clamp configuration: 145 mM K-gluconate, 5 mM HEPES, 2 mM EGTA, 2 mM Mg-gluconate (0.1 mM free Mg^{2+}), and 2 mM K$_2$ATP; adjust to pH 7.2 with KOH. Because this solution, here called solution a, creates a large junction potential drift when in direct contact with the Ag/AgCl electrode, a second pipet solution containing chloride, solution b, is carefully added to fill the upper part of the pipet bathing the intrapipet electrode. Solution b contains 145 mM KCl, 5 mM HEPES, and 5 mM EGTA, with adjustment to pH 7.2 with KOH.
4. Pipet (intracellular) medium for the permeabilized patch configuration: Solution a of 145 mM K-gluconate, 1 mM EGTA, and 10 mM HEPES and 0.2–0.4 µg/mL amphotericin B is added extemporarily and medium is brought to pH 7.3 with KOH; solution b is 145 mM KCl, 5 mM HEPES, and 5 mM EGTA brought to pH 7.2 with KOH.
5. Standard extracellular and pipet solution b medium for the giant patch configuration: 140 mM KCl, 10 mM HEPES, and 1 mM EGTA adjusted to pH 7.3 with KOH.
6. Microperfusion and pipet solution a medium for the giant patch configuration: 145 mM K-gluconate, 10 mM HEPES, and 1 mM EGTA adjusted to pH 7.3 with KOH.
7. Borosilicate glass capillaries (glass type 8250, Garner Glass, Claremont, CA) for giant patch configuration.
8. Soda lime glass nonheparinized microhematocrit capillaries (Kimble, Vineland, NJ) for other patch clamp configurations.
9. Attofluor® cell chamber (Molecular Probes, Eugene, OR).
10. The conventional material needed for patch clamp has been described *(12,13)*. The Eclipse TS100 inverted microscope (Nikon, Tokyo, Japan) is equipped with halogen epifluorescence for GFP visualization. The custom-made 37°C-heated chamber and microperfusion system are shown in **Fig. 1**. The Tygon capillaries (0.25-mm i.d.) used for the microperfusion system are glued together. For giant

Fig. 1. Schematic drawing of the cell chamber used for patch clamp. The bath electrode and the general perfusion (1 mL/min) inlet and outlet are not shown for clarity.

patch experiments, an excision pipet has been added to suck and excise the cell after the seal formation (**Fig. 1**).

3. Methods

Because KCNQ1-related channelopathies may be the consequence of different molecular alterations, various investigations should be conducted.

3.1. How to Evaluate Protein Processing

Mutation in any member of the channel complex may lead to alteration of the protein synthesis, stability, or trafficking. Using COS-7 cells as host for heterologous expression, trafficking of the different partners of the complex is investigated.

3.1.1. Cell Culture

1. COS-7 cells are passaged twice a week on 25-cm^2 flasks, and 1:25 to 1:40 splits of the cells are used before they reach confluence.
2. For Western blotting, 25-cm^2 flasks are seeded for transfection.
3. For protein localization, cells are passaged on 35-mm Petri dishes bottomed with glass cover slips. A 1:20 split of the cells gives 60–70% confluence 24 h after splitting, which is required for optimal transfection efficacy.

3.1.2. Transfection

For transfection using Fugene-6 or Jet-PEI, cells are transfected according to the standard protocol recommended by the manufacturer. Both reagents are equally easy to use and efficient. Transfection is used preferentially when a large number of transfected cells are needed, such as for Western blot (*6*) or immunocytochemistry experiments.

1. The culture medium (with supplements as described in **Subheading 2.1.**, **item 2**) of the flasks or Petri dishes is renewed before transfection.
2. For a flask, 10 µL of Jet-PEI diluted in 125 µL plasmid buffer are complexed with 5 µg DNA diluted in 125 µL plasmid buffer for 30–45 min before addition to the flask.

3. For a 35-mm Petri dish and JetPEI, the amounts are reduced to transfect 2 μg DNA (4 μL Jet-PEI and 50 μL plasmid buffer). For Fugene-6, 6 μL reagent diluted in 100 μL Dulbecco's modified Eagle's medium are complexed with 2 μg DNA diluted in 10 μL plasmid buffer for 30–45 min before addition to the dish.

3.1.3. Western Blot

1. At 24 h post-transfection, cells are washed twice with cold PBS, incubated with 200 μL SDS gel-loading buffer, and scraped.
2. The samples are placed in a boiling water bath for 10 min, and the chromosomal DNA is sheared by sonication for 2 min.
3. Samples are centrifuged at 10,000g for 10 min at room temperature, and the supernatant is used for immunoblotting.
4. Protein concentration is determined using the Bradford assay (bovine serum albumin as a standard). Proteins (15 μg/well) are separated by electrophoresis on 7.5% polyacrylamide gels using the Laemmli system.
5. The separated proteins are electrotransferred to an Immobilon-P polyvinylidene difluoride membrane (Sigma). Electroblotting is carried out in Tris-buffer containing 40.7 mM glycine and 20% (v/v) methanol at pH 8.3 and 100 V for 1.5 h.
6. For immunodetection of the proteins, blots are blocked with PBS supplemented with 5% (w/v) nonfat dry milk and 0.1% Tween-20 for 1 h at 37°C.
7. The monoclonal anti-HA antibody is diluted 500-fold in PBS containing 2.5% (w/v) nonfat dry milk and 0.1% (v/v) Tween-20 and incubated with the blots for 1 h at room temperature.
8. The filters are rinsed five times with quench buffer and incubated for 1 h at room temperature in a 1:1000 dilution of the biotin-conjugated goat antimouse antibody.
9. The blots are then incubated for 30 min with conjugated extravidin peroxidase and finally developed by the enhanced chemoluminescence.
10. Western blotting system: Biomax-MR Kodak films are used to detect protein in the blot.

Using this approach, we have shown that the alternative splice short KCNQ1 isoform is fully processed in COS-7 cells transfected with HA-tagged plasmids *(6)*. A protein with a 61-kDa molecular mass as predicted by the coding sequence is detected. A second band is visible on the immunoblot, which likely corresponds to the unglycosylated form of the protein.

3.1.4. Fluorescent-Tagged Proteins

Fluorescent tags are used to localize proteins in the different cell compartments.

1. At 48 h after transfection, cells transfected with tagged proteins are washed three times with PBS and fixed for a few minutes with 4% (v/v) paraformaldehyde in PBS.
2. Cells are mounted in mowiol 4-88 (Calbiochem) after rinsing three times with PBS. The cells are observed with a scanning laser confocal microscope using an oil immersion ×100 lens.

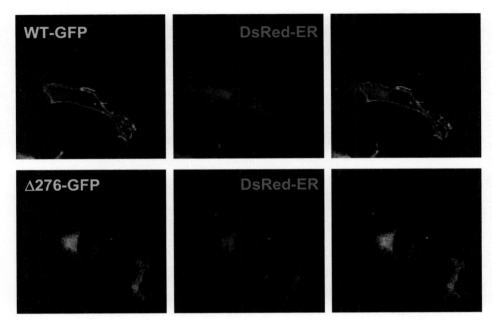

Fig. 2. Alteration of the mutant channel trafficking. Confocal laser scanning imaging of COS-7 cells transfected with wild-type-green fluorescent protein (WT-GFP) (WT-GFP, top) or Δ276 (Δ276-GFP, bottom) GFP-tagged KCNQ1 and DsRed-ER. **Left**: Localization of GFP-tagged KCNQ1 (green). **Middle**: Localization of DsRed-ER (red). **Right**: Superimposition of both images. (Courtesy of F. Potet.)

Typical observations of fluorescent-tagged proteins are shown in **Fig. 2**. When expressed alone, WT GFP-KCNQ1 localizes at the cell membrane, whereas Romano-Ward syndrome-associated ΔS276 GFP-KCNQ1 colocalizes at the endoplasmic reticulum with the DsRed-KDEL sequence of calreticulin *(17)*.

3.2. Evaluation of Altered Mutated Channel Function

Mutation in the protein constituting the channel pore (i.e., KCNQ1 isoform 1) can lead to proteins fully processed at the cell membrane. Function alteration thus can be evaluated using the patch clamp technique. For measurement of the current amplitude, expression of the mutant in the presence of KCNE1 can be achieved in transfected cells using different patch clamp configurations.

3.2.1. Cell Culture and Transfection

1. Cells are passaged on 35-mm Petri dishes and transfected using Fugene-6 or JetPEI (*see* **Subheadings 3.1.1.** and **3.1.2.**).

2. To limit current amplitude to conditions compatible with proper membrane voltage control (*see* **Note 2**), only 20% of the transfected plasmids encode for the channel protein. The transfecting plasmid mixture thus contains 20% WT or mutant KCNQ1 isoform 1 and 20–40% KCNE1 completed with 40–60% GFP coding plasmid.
3. At 8–24 h after cell transfection, cells are passaged again to avoid confluence. This is critical for macroscopic current recordings, which require isolated cells for proper voltage control. Cells are seeded on 25-mm diameter glass cover slips placed in 35-mm Petri dishes. Glass cover slips optimize the GFP signal. This is especially needed when a small amount of GFP-expressing plasmid is used.

3.2.2. Patch Clamp: Whole-Cell Configuration

The patch clamp whole-cell configuration is used for fast screening of the mutants for their functional alterations. Current recordings are performed 1 min after rupturing the patch. After this delay, KCNQ1 currents (in the absence or presence of KCNE1) undergo rapid decay (rundown) caused by dilution of various cytosol components into the pipet.

1. Patch pipets are pulled with a P-30 puller (Sutter Instruments, Novato) to obtain pipet resistance of 2–3 MΩ with the pipet solution used.
2. The cells on the cover slip are mounted on the Attofluor cell chamber. The chamber is then placed in the heating device (**Fig. 1**) on the microscope stage. Experiment temperature is 35°C.
3. To eliminate the endogenous chloride currents present in COS-7 cells, patch pipets are filled with the chloride-free solution a. The electrical contact between the pipet solution and the Ag/AgCl electrode is made through solution b containing chloride.
4. A green GFP-positive cell is selected using the epifluorescence light of the microscope (we use filters for fluorescein isothiocyanate to detect the GFP fluorescence).
5. The protocol to seal the cell and rupture the patch is standard (*12*).
6. It is most important to determine current characteristics immediately after patch rupture before dilution of the cytosol in the pipet.

A typical recording of I_{Ks} obtained in a cell transfected with 20% KCNQ1 isoform 1, 40% KCNE1, and 40% GFP is shown in **Fig. 3A,B**.

3.2.3. Patch Clamp: Permeabilized Patch Configuration

A mutation may have an impact on kinetics and voltage dependency of the mutated channel. The permeabilized patch configuration (also called perforated patch configuration) is used as an alternative to the ruptured patch because the former allows complex stimulation protocols in the absence of rundown of the channel activity.

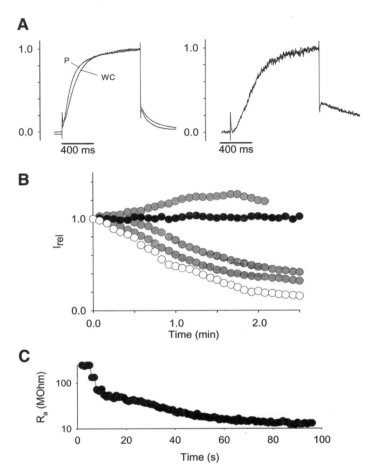

Fig. 3. Representative recordings obtained with different patch clamp configurations. (**A**) **Left**: representative recordings of the KCNE1-KCNQ1 current in whole-cell (WC) and permeabilized patch (P) configurations at 35°C. The voltage protocol consists of depolarized voltage steps from a holding potential (–80 mV) to +40 mV (1 s) and then back to –40 mV (500 ms) every 5 s. The traces have been normalized to compare activation and deactivation kinetics. **Right**: representative recording of the KCNE1-KCNQ1 current in giant patch configurations at 22°C. (**B**) Representative evolution of the KCNE1-KCNQ1 current amplitude with time. Currents were normalized to compare rundown kinetics. Gray circles stand for three different cells in the whole-cell configuration and illustrate the variability of the rundown under this condition. Black circles stand for a representative cell in the permeabilized patch configuration. Open circles stand for a representative cell in the giant patch configuration in the absence of PIP_2 and MgATP. (**C**) A representative recording of the evolution of the access resistance (R_a) during amphotericin B partitioning.

1. *See* **steps 1–4** in **Subheading 3.2.2.**
2. After the seal is achieved, the capacitive current provoked by a voltage step is measured to calculate the access resistance. We use a voltage step from –60 to –50 mV for 50 ms.
3. Amphotericin B (*see* **Subheading 2.5.**, **step 4**) partitioning in the patch membrane leads to a gradual decrease in the access resistance (**Fig. 3C**). Macroscopic currents are recorded when the access resistance is below 10 MΩ.
4. Because the access resistance creates a shift between theoretical and true membrane potential (*see* **Note 2**), especially when large currents are recorded, compensation for access resistance should be conducted. The access resistance should also be determined after current acquisition to check for possible variation.

A typical recording of I_{Ks} obtained in a cell transfected with 20% KCNQ1 isoform 1, 40% KCNE1, and 40% GFP is shown in **Fig. 3A,B**. Note that current activation and deactivation kinetics are not modified depending on the patch clamp configuration (**Fig. 3A**), whereas no rundown is observed when the permeabilized patch is used (**Fig. 3B**). This configuration allowed us to characterize the alterations observed with the short QT-linked mutation V307L of KCNQ1 *(10)*. Acceleration of the activation kinetics associated with a shift of the half-activation potential was observed. When introduced in a human action potential computer model, the modified biophysical parameters leading to a gain of function predicted repolarization shortening.

3.3. Evaluation of a Dominant Negative Activity

As mentioned, the alternative isoform 2 of KCNQ1 exerts a pronounced negative dominance on KCNQ1 isoform 1 channels. Furthermore, isoform 1 KCNQ1 mutants may have dominant negative activity toward the WT channel protein. Dominant negative activity can be observed by means of the patch clamp, and its mechanism may be detailed using immunocytochemistry.

3.3.1. Functional Alteration: Cell Culture and Intranuclear Microinjection

A dominant negative activity is revealed when coexpression of the alternative isoform or mutant channel induces the reduction of the WT channel activity. Accurate current amplitude measurement is thus needed under conditions by which heterologous proteins are expressed at a precise and constant ratio. Plasmid microinjection is preferred to transfection to reach this goal.

1. For intranuclear microinjection, cells are seeded at very low density (about 1:60) on 35-mm Petri dishes bottomed with cover slips. Cells are seeded 24 h before microinjection.
2. Plasmids are diluted at a concentration of 5–50 μg/mL in the plasmid buffer supplemented with 0.5% 150-kDa fluorescein isothiocyanate-dextran (Sigma).

3. Plasmids are filtered on VectaSpin microfuge filters at 11,000g for 15 min at room temperature to prevent clogging of the pipet tip.
4. The Petri dishes are filled with 3 mL medium to prevent visible light diffraction at the pipet-solution surface interface. Diffraction alters the phase contrast view of the cell.
5. Intranuclear microinjection is performed using the Z-limit option with a 0.3-s injection time and 30–80 hPa injection pressure (*see* **Note 3**). We routinely inject around 100 nuclei in less than 30 min.
6. Unless the microinjection equipment is placed into a cell culture cabinet, the room in which injection is performed should be dedicated to intranuclear injection to limit microbial contamination because microinjections are not performed under sterile conditions.

3.3.2. Traffic Alteration: Intracellular Localization

Immunoreactive or fluorescent tags can be used to localize proteins in the different cellular compartments and to assess the mechanisms leading to a dominant negative effect (e.g., retention in the endoplasmic reticulum). Dominant negative mechanisms of isoform 2 were revealed using this approach (**Fig. 4**).

1. Cells are cultured on cover slips and transfected as in **Subheading 3.1.**
2. At 48 h after transfection, cells transfected with HA-tagged channel plasmids are washed three times with PBS and fixed with 4% (v/v) paraformaldehyde in PBS. They are then incubated with the monoclonal anti-HA antibody (1:1000) in PBS with 1% ovalbumin for 48 h at 4°C. Under this condition, the cells are not permeabilized, and only epitopes present at the membrane surface are detected.
3. Alternatively, cells are permeabilized by addition of 0.1% Tween-20 to detect intracellular retained proteins.
4. Cells are washed three times and incubated for 1 h at room temperature with the tetramethylrhodamine isothiocyanate-conjugated goat antirabbit immunoglobulin G (1:400) in PBS plus 1% ovalbumin.
5. For HA-tagged or fluorescent-tagged channels (cf. **Subheading 3.1.4.**), cells are mounted in mowiol 4-88 (Calbiochem) after rinsing three times with PBS and observed with a scanning laser confocal microscope using an oil immersion ×100 lens.

The use of external HA tags also discriminates the WT KCNQ1 channel fraction present at the cell membrane (WT-HA) from that retained by mutated GFP-tagged channel mutants as illustrated in **Fig. 5**. Romano-Ward syndrome-associated Y315S KCNQ1 channel prevented WT KCNQ1 channels trafficking to the membrane, whereas Jervell and Lange-Nielsen syndrome-associated Δ544 or R243H KCNQ1 did not. Mutations may impair trafficking (as Y315S or Δ544) or the mutant channel function only (as R243H *[18]*).

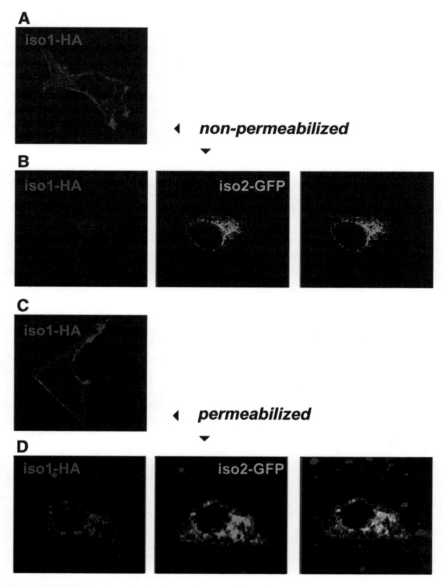

Fig. 4. KCNQ1 isoform 2 alters I_{Ks} by retaining KCNQ1 isoform 1 in intracellular compartments. Confocal laser scanning imaging of COS-7 cells transfected with HA-tagged KCNQ1 isoform 1 (iso1-HA, **A** and **C**) or HA-tagged KCNQ1 isoform 1 and green fluorescent protein (GFP)-tagged KCNQ1 isoform 2 (iso2-GFP) in non-permeabilized (**B**) and permeabilized (**D**) conditions. The iso1-HA is visualized (**left**) in nonpermeabilized cells when expressed at the membrane level only. Its colocalization with iso2-GFP is visualized in permeabilized cells (**D**, right). (Courtesy of F. Potet and S. Alcolea.)

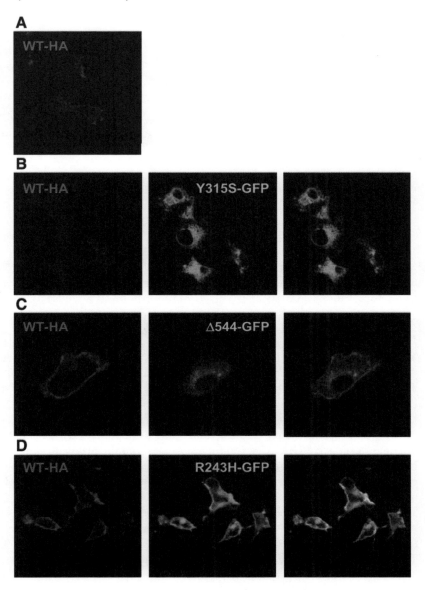

Fig. 5. Alterations on KCNQ1 isoform 1 trafficking as induced by Romano-Ward (Y315S) or Jervell and Lange-Nielsen (Δ544 and R243H) mutations. Confocal laser scanning imaging of nonpermeabilized COS-7 cells transfected with HA-tagged wild-type (WT) KCNQ1 channel (WT-HA) alone (**A**) or with green fluorescent protein (GFP)-tagged mutant KCNQ1 channels (**B–D**). **Left**: fraction of HA-tagged WT KCNQ1 present at the membrane level. **Middle**: localization of the GFP-tagged mutant KCNQ1. **Right**: superimposition of both images. (Courtesy of F. Potet and S. Alcolea.)

3.4. Evaluation of Altered Channel Regulation

Mutations of KCNQ1 channels may alter their regulation by intracellular agents. We have shown that phosphatidylinositol 4,5-bisphosphate (PIP$_2$) and intracellular MgATP (adenosine triphosphate) control the activity of the KCNQ1-KCNE1 potassium channel complex *(19)*. We have adapted the giant patch configuration developed by Hilgemmann et al. in large cells such as cardiomyocytes *(20)* or *Xenopus* oocytes *(21)* and more recently in smaller cells *(22,23)*.

3.4.1. Cell Culture and Transfection

Cells are cultured and transfected as in **Subheading 3.1.**

3.4.2. Patch Clamp: Giant Patch Configuration

The giant patch configuration is similar to the widely used inside-out configuration, with a main difference in the size of the patch. Because KCNE1-KCNQ1 unitary conductance is small (less than a picoSiemens), the patch should contain a large number of channels to record sizeable KCNE1-KCNQ1 current. The use of *Xenopus* oocytes to express and study human KCNQ1 is limited by endogenous expression of *Xenopus* KCNQ1.

1. The cells on the cover slip are mounted on the Attofluor cell chamber. Experiment temperature is generally 22°C to improve patch stability.
2. The patch pipet is pulled from glass capillaries (cf. **Subheading 2.5.**, **step 7**) on a P-30 puller (Sutter Instruments) to make a long and fine tip (less than 1 μm).
3. Using a microforge MF-83 (Narishige), the patch pipet tip is melted against the microforge filament coated with the same glass as the pipet (Garner 8250) and the microforge HEATER ADJ knob is set to 20 (with the original filament; *see* **Note 4**). Then, the microforge filament is switched off, and the pipet is quickly pulled back to obtain a diameter of 15–21 μm (five to seven graduations on the microforge reticule).
4. The pipet is polished close to the filament (5–10 μm); the filament is lit (HEATER ADJ. knob set to 30; *see* **Note 4**). The final diameter should be 9–12 μm (three to four graduations on the microforge reticule).
5. The pipet is filled with the extracellular solution, installed on the electrode holder, and connected to a 1-mL syringe to perform the seal by suction. It is also possible to use an air-filled microsyringe (Captrol III, Drummond Scientific, Broomall, PA; *23*).
6. The excision pipet is pulled and polished just like the patch pipet, although it is slightly larger (approx 15 μm). It is connected to a 10-mL syringe.
7. Cells are placed in the giant patch extracellular medium for 15–30 min. In the absence of divalent cations and the presence of a high potassium concentration, the cells gradually round.

8. After selection of the cell (usually with GFP fluorescence), the seal is obtained by approaching the pipet close to the cell under the microscope (×400) and applying gentle suction (maximum 100 μL). Then, the giant patch configuration is obtained by sucking the cell out with the excision pipet.

Typical recordings are shown in **Fig. 3A,B**. Using the same configuration, we have shown that PIP$_2$ affinity is reduced in three KCNQ1 mutant channels (R243H, R539W, R555C) associated with the long QT syndrome *(24)*.

4. Notes

1. If the biophysical characteristics of the channel are altered ($V_{0.5}$, activation and deactivation kinetics), then there is a chance that the single-channel open probability (Po) is modified. Because the macroscopic current depends not only on the channel number but also on the Po ($I = n*Po*\gamma*Vm$, with γ the single-channel conductance and Vm the membrane potential), the macroscopic current amplitude is not a good indicator of the channel number. To use the macroscopic current amplitude as an indicator of the channel number, be sure that the channel conductance and open probability are not altered.
2. Because $U = RI$, if the current measured is 1 nA and the access resistance is 20 MΩ, then the shift in the membrane potential is 20 mV.
3. Do not hesitate to replace the pipet regularly because pipets erode after a few tens of injections.
4. The HEATER ADJ. may be different from one microforge to another. The first setting (20 in our case) should melt the pipet only when in contact. The second setting should melt the pipet in seconds when the pipet and the filament are 5–10 μm apart.

References

1. Sanguinetti, M. C., Curran, M. E., Zou, A., et al. (1996) Coassembly of KvLQT1 and minK (IsK) proteins to form cardiac I_{Ks} potassium channel. *Nature* **384,** 80–83.
2. Barhanin, J., Lesage, F., Guillemare, E., Fink, M., Lazdunski, M., and Romey, G. (1996) KvLQT1 and lsK (minK) proteins associate to form the $I\,I_{Ks}$ cardiac potassium current. *Nature* **384,** 78–80.
3. Marx, S. O., Kurokawa, J., Reiken, S., et al. (2002) Requirement of a macromolecular signaling complex for beta adrenergic receptor modulation of the KCNQ1-KCNE1 potassium channel. *Science* **295,** 496–499.
4. Furukawa, T., Ono, Y., Tsuchiya, H., et al. (2001) Specific interaction of the potassium channel β-subunit minK with the sarcomeric protein T-cap suggests a T-tubule-myofibril linking system. *J. Mol. Biol.* **313,** 775–784.
5. Potet, F., Scott, J. D., Mohammad-Panah, R., Escande, D., and Baró, I. (2001) AKAP proteins anchor cAMP-dependent protein kinase to KvLQT1/IsK channel complex. *Am. J. Physiol. Heart. Circ. Physiol.* **280,** H2038–H2045.

6. Demolombe, S., Baró, I., Péréon, Y., et al. (1998) A dominant negative isoform of the long QT syndrome 1 gene product. *J. Biol. Chem.* **273,** 6837–6843.

7. Péréon, Y., Demolombe, S., Baró, I., Drouin, E., Charpentier, F., and Escande, D. (2000) Differential expression of KvLQT1 isoforms across the human ventricular wall. *Am. J. Physiol. Heart Circ. Physiol.* **278,** H1908–H1915.

8. Roden, D. M., Lazzara, R., Rosen, M., Schwartz, P. J., Towbin, J. , and Vincent, G. M. (1996) Multiple mechanisms in the long-QT syndrome. *Circulation* **94,** 1996–2012.

9. Neyroud, N., Tesson, F., Denjoy, I., et al. (1997) A novel mutation in the K⁺ channel KvLQT1 causes the Jervell and Lange-Nielsen cardioauditory syndrome. *Nat. Genet.* **15,** 186–189.

10. Bellocq, C., van Ginneken, A. C., Bezzina, C. R., et al. (2004) Mutation in the KCNQ1 gene leading to the short QT-interval syndrome. *Circulation* **109,** 2394–2397.

11. Chen, Y. H., Xu, S. J., Bendahhou, S., et al. (2003) KCNQ1 gain-of-function mutation in familial atrial fibrillation. *Science* **299,** 251–254.

12. Boulton, A. A., Baker, G. B., and Walz, W. (eds.). (1995) *Patch-Clamp Applications and Protocols*, Humana Press, Totowa, NJ.

13. Levis, R. A. and Rae, J. L. (1992) Constructing a patch clamp setup. *Methods Enzymol.* **207,** 14–66.

14. Yu, S. P. and Kerchner, G. A. (1998) Endogenous voltage-gated potassium channels in human embryonic kidney (HEK293) cells. *J. Neurosci. Res.* **52,** 612–617.

15. Chouabe, C., Neyroud, N., Guicheney, P., Lazdunski, M., Romey, G., and Barhanin, J. (1997) Properties of KvLQT1 K⁺ channel mutations in Romano-Ward and Jervell and Lange-Nielsen inherited cardiac arrhythmias. *EMBO J.* **16,** 5472–5479.

16. Ottschytsch, N., Raes, A., van Hoorick, D., and Snyders, D. J. (2002) Obligatory heterotetramerization of three previously uncharacterized Kv channel α-subunits identified in the human genome. *Proc. Natl. Acad. Sci. USA* **99,** 7986–7991.

17. Gouas, L., Bellocq, C., Berthet, M., et al. (2004) New KCNQ1 mutations leading to haploinsufficiency in a general population; defective trafficking of a KvLQT1 mutant. *Cardiovasc. Res.* **63,** 60–68.

18. Chouabe, C., Neyroud, N., Richard, P., et al. (2000) Novel mutations in KvLQT1 that affect I_{Ks} activation through interactions with IsK. *Cardiovasc. Res.* **45,** 971–980.

19. Loussouarn, G., Park, K.-H., Bellocq, C., Baró, I., Charpentier, F., and Escande, D. (2003) Phosphatidylinositol-4,5-bisphosphate, PIP_2, controls KCNQ1/KCNE1 voltage-gated potassium channels: a functional homology between voltage-gated and inward rectifier K⁺ channels. *EMBO J.* **22,** 5412–5421.

20. Hilgemann, D. W. (1989) Giant excised cardiac sarcolemmal membrane patches: sodium and sodium-calcium exchange currents. *Pflugers Arch.* **415,** 247–249.

21. Hilgemann, D. W. (1994) Channel-like function of the Na,K pump probed at microsecond resolution in giant membrane patches. *Science* **263,** 1429–1432.

22. Hilgemann, D. W. and Lu, C. C. (1998) Giant membrane patches: improvements and applications. *Methods Enzymol.* **293,** 267–280.
23. Couey, J. J., Ryan, D. P., Glover, J. T., Dreixler, J. C., Young, J. B., and Houamed, K. M. (2002) Giant excised patch recordings of recombinant ion channel currents expressed in mammalian cells. *Neurosci. Lett.* **329,** 17–20.
24. Park, K.-H., Piron, J., Dahimene, D., et al (2005). Impaired KCNQ1/KCNE1 and phosphatidylinositol-4,5-bisphosphate Interaction underlies the long QT syndrome. *Circ. Res.* **96,** 730–739.

15

Tissue-Specific Transgenic and Knockout Mice

Andrée Porret, Anne-Marie Mérillat, Sabrina Guichard, Friedrich Beermann, and Edith Hummler

Summary

Analysis of genetically engineered mice is crucial for our understanding of the in vivo function of genes and proteins in the whole organism. This includes inactivation of a gene or the generation of specific mutations. The development of knockout and transgenic technologies in the mouse, therefore, represents a powerful tool for elucidating gene function, for modeling of human diseases, and potentially for the evaluation of drugs. In particular, conditional gene targeting applying the Cre/loxP-mediated recombination system is increasingly used to evaluate the role of the gene of interest in a cell-type-specific or even inducible manner. The experimental steps start with the characterization of the gene locus, followed by construction of a vector, gene targeting in ES cells, and establishment of mouse lines carrying the desired mutation. These are then bred to transgenic mice expressing Cre recombinase in a tissue-specific manner, thus allowing gene inactivation in a cell type of interest.

Key Words: Blastocyst injection; conditional; Cre recombinase; ENaC; ES cell culture; gene targeting; homologous recombination; knockout; mouse; *Scnn1a*; transgenic.

1. Introduction

The generation of mutant mice via gene targeting has become important in determining the function of specific genes/proteins in vivo. Conditional gene targeting using the Cre/loxP system furthermore allows spatial and temporal control of gene inactivation in the mouse. To produce a mutant mouse strain by gene targeting and homologous recombination in embryonic stem (ES) cells, a construct is prepared that contains the desired mutation. This targeting construct is then transferred into ES cells that are derived from the inner cell mass of a blastocyst-stage embryo. Homologous recombination occurs in a small

From: *Methods in Molecular Biology, vol. 337: Ion Channels: Methods and Protocols*
Edited by: J. D. Stockand and M. S. Shapiro © Humana Press Inc., Totowa, NJ

number of the transfected cells and results in introduction of the mutation into the target gene. Once identified, mutant ES cell clones are injected into blastocysts to produce chimeric mice.

ES cell lines retain the ability to differentiate into every cell type present in the mouse, including the germline. Thus, breeding of chimeras yields animals that carry the mutant ES cell genome and can then be bred to generate mice homozygous mutant for the modified allele. These are further bred to Cre-recombinase transgenic mice to remove the gene/protein of interest everywhere (null allele) or in a specific organ.

1.1. Conditional Knockouts

Using conditional gene-targeting strategy, starting from one single floxed mouse strain, several mutants can be generated, ranging from the complete knockout to more defined mutations in single organs. This strategy exploits the Cre-loxP system. The *Cre* gene of the bacteriophage P1 encodes a site-specific recombinase that efficiently excises intervening DNA sequences located between two loxP sites in the same orientation, leaving one loxP site on the DNA *(1,2)*. Thus, the system can be used to generate specific genome alterations such as deletions, insertions, translocations, or inversions, depending on the location and orientation of the loxP sites *(3)*.

In a conditional knockout approach, coding sequences essential for gene function (vital region) are flanked by two loxP sites. Cre-mediated deletion of these sequences should convert the floxed allele into a null allele. The selection marker neo is flanked by an additional loxP site (*see* **Subheading 3.1.1.**) or, alternatively, by frt sites ("flirted") that can then be deleted using Cre and Flp recombinases. This step is either performed in vitro in targeted ES cells by transient expression of eukaryotic FLPe or Cre expression vectors *(4,5)* or in vivo using, for instance, transgenic mice expressing the recombinase to finally obtain offspring without the selection marker.

In this chapter, we describe relevant steps for developing mice harboring a conditional allele. As an example, we discuss the conditional knockout of the α-subunit of the epithelial Na$^+$ channel (ENaC) (*Scnn1a^{flox}/Scnn1a^{flox}*) *(6)*. ENaC is a heteromultimeric protein made up of three homologous subunits (α, β, and γ) encoded by three different genes (*Scnn1a, Scnn1b,* and *Scnn1g,* respectively) localized on chromosomes 6 and 7 in the mouse *(7)*. Complete inactivation of all three *Scnn1* genes (encoding for α, β, and γ ENaC) leads to early postnatal death *(8–10)*. To address (α) ENaC deficiency in selected tissues, we therefore decided to generate a conditional allele at the gene locus encoding the α subunit (*Scnn1a*) *(6)*. Mice carrying such a targeted allele are now used to study ENaC absence in a cell-type-specific or time-dependent manner *(11)*. The gene is not affected in other cells/organs that show wild-type expression levels.

2. Materials

All solutions should be prepared in high-quality, molecular-biology-grade water (e.g., MilliQ H_2O). All solutions used for ES cell culture and embryo handling must be made up in embryo-grade pure water, and tissue-culture-grade plasticware is preferable to avoid detergents, which are toxic to embryonic cells. Unless otherwise specified, standard biological suppliers are used for enzymes and chemicals.

2.1. Cloning of the Targeting Construct

1. *Escherichia coli* strains (e.g., DH5α, HB101).
2. Luria-Bertani (LB) medium, LB agar plates.
3. Ampicillin, IPTG (isopropyl-β-D-thio-galactopyranoside).
4. Ligase and ligase buffer (Roche, rapid DNA ligation kit 1635379).
5. 100% isopropanol and 100% ethanol.
6. TE buffer: 10 mM Tris-HCl at pH 7.5) and 1 mM ethylenediaminetetraacetic acid (EDTA) (8.0) pH 7.6.
7. Oligonucleotide primers, 100 mM deoxyribonucleotide 5'-triphosphates (dNTPs) for polymerase chain reaction (PCR) (Amersham, sequencing grade 27-2035-01).
8. Restriction enzymes with 10X reaction buffers (Invitrogen).
9. Taq polymerase (Amersham 27-0799-06) and Klenow polymerase (Roche) with corresponding 10X buffers.
10. Phenol/chloroform/isoamyl alcohol 25:24:1.
11. DNA-grade agarose (Eurogentec EP-0010-05) and DNA sequencing gel equipment.
12. DNA electrophoresis equipment (WitecAg, Switzerland, Southern blot cooled Maxi Plus, CHU25).
13. 1X TBE buffer: 89 mM Tris-borate and 2.5 mM EDTA.
14. 1X TAE buffer: 40 mM Tris-acetate and 1 mM EDTA.
15. Dye loading buffer (TBE gels: "blue juice," final concentration): 33% glycerol, 10 mM Tris-HCl (pH 8.0), 20 mM EDTA (pH 8.0), 0.04% bromophenol blue, 0.04% xylene cyanol FF (optional).
16. Dye loading buffer (TAE gels): 60% sucrose, 0.05% bromophenol blue, and 40 mM EDTA (pH 8.0).
17. DNA size marker (e.g., 1-kb DNA ladder, Invitrogen, cat. no. 15615-024).
18. Qiagen Plasmid Mini (27106) and Midi purification kit (12643).
19. Qiagen PCR purification and gel extraction kit (28706).
20. Mouse bacterial artificial chromosome (BAC) library (e.g., 129SvEvTac, RPCI-22, http://bacpac.chori.org) or BAC clones (Sanger center, http://www.ensembl.org/Mus_musculus/).

2.2. ES Cell Culture and Analyses

1. Dulbecco's modified Eagle's medium (DMEM; 4.5 g/L glucose, pyruvate, Glutamax-1, Gibco, cat. no. 31966-021).
2. Ultrapure water (Gibco, cat. no. 15230-089).

3. 0.25% Trypsin-EDTA (Gibco, cat. no. 25200-056).
4. Penicillin/streptomycin (Gibco, cat. no. 15070-063).
5. 1 M HEPES (Gibco, cat. no. 15630-056).
6. 50 mM β-mercaptoethanol (Gibco, cat. no. 31350-010).
7. Phosphate-buffered saline (PBS) without Ca^{2+}/Mg^{2+} (Gibco, cat. no. 10010-015).
8. Fetal calf serum (FCS; Hyclone, cat. no. CH 30160.03).
9. Leukemia inhibitory factor (ESGRO Chemicon ESG, cat. no. 1107).
10. Geneticin G418 (Gibco, cat. no. 11811-023).
11. Gancyclovir (Roche Cymevene 500 mg).
12. Dimethyl sulfoxide (DMSO; Merck, cat. no. 1.09678.0100).
13. Gelatin from porcine skin type A (Sigma, cat. no. G1890): 0.1% solution in double-distilled water (ddH$_2$O), autoclave, and filter.
14. 37–60°C Incubator (Binder, B28, cat. no. 9110-0004).
15. Inverted microscope (e.g., Nikon Eclipse TS100, with objectives ×4, ×10, ×20).
16. Tissue culture plasticware: 96-well plates (flat bottom, Falcon, cat. no. 35 3072); 96-well plates (V bottom, Nunc, cat. no. 249662); microwell lid (Nunc, cat. no. 264122); 24-well plates (Falcon, cat. no. 353047); 6-well plates (Falcon, cat. no. 353046); 6-cm dishes (Falcon, cat. no. 353004); 10-cm dishes (Falcon, cat. no. 353003); 25-cm^2 (Falcon, cat. no. 353014) and 75-cm^2 (Falcon, cat. no. 353135).
17. Electroporation apparatus (e.g., BTX Square Wave Electroporation System) and equipment (0.4-cm cuvets; BTX 640).
18. Lysis buffer: 10 mM Tris-HCl at pH 7.5, 10 mM EDTA, 10 mM NaCl, 0.5% N-laurylsarcosine, and proteinase K (0.5 mg/mL).
19. 5 M NaCl.
20. 99% EtOH.
21. Restriction enzyme digestion mixture: 1X restriction buffer, 1 mM spermidine (Sigma, cat. no. 50266-1G), and 40 U each restriction enzyme.
22. 50X TAE: 242 g Tris-base, 57.1 mL acetic acid, and 100 mL 0.5 M EDTA (pH 8.0); add to 1 L ddH$_2$O.
23. Phosphate buffer: 89 g Na$_2$HPO$_4$, adjust to pH 7.2 with H$_3$PO$_4$ (phosphoric acid, about 3 mL), and add to 1 L ddH$_2$O.
24. Neutralizing solution (Southern blot analysis): 1.5 M NaCl, 0.5 M Tris-HCl (pH 7.2), and 1 mM EDTA (pH 8.0).
25. Prehybridization/hybridization solution (Southern blot analysis): 7% sodium dodecyl sulfate (SDS; Fluka, cat. no. 71729), 0.5 M phosphate buffer, 0.5 M EDTA, and 0.25% milk powder (e.g., Migros Switzerland).
26. Washing solution (Southern blot analysis): 0.04 M phosphate buffer (pH 7.2), 1% SDS, and 5 mM EDTA (pH 8.0).
27. TaKaRa La Taq DNA polymerase (Takara Bio Inc., cat. no. RR002A).
28. Colchicin (Sigma Demecolcin solution D1925).
29. Giemsa (Fluka, cat. no. 48900).
30. Proteinase K (Merck, cat. no. 1.24568.0100).
31. Styrofoam box or freezing container (Nunc Nalgene, cat. no. 5100-0001).
32. Blotting membranes (Amersham Hybond-N, cat. no. RPN303 N).

2.3. Embryo Handling and Manipulations

1. Pregnant mare's serum gonadotropin (Sigma, cat. no. G4877, 10,000 IU) and human chorionic gonadotropin (Sigma CG 10, 10,000 IU) are diluted in PBS, aliquoted in Eppendorf tubes at 50 IU (100–200 μL), and stored at –20°C. Before use, PBS is added to 1.5 mL, and 150 μL (= 5 IU) are injected intraperitoneally per mouse.
2. M2 medium (Sigma, cat. no. M7167).
3. M16 medium (Sigma, cat. no. M7292).
4. KSOM medium (Specialty Media, cat. no. MR-020P-F).
5. Dow Corning 200/50 cS fluid (BHD, VWR, cat. no. 630064V).
6. Embryo-tested mineral oil (Sigma, cat. no. M8410).
7. Hyaluronidase type IV-S from bovine testes (Sigma H4272 or H3384, 10 mg/mL in water).
8. Disposable 1-mL syringes (23-, 27-, or 30-gage needle).
9. Microinjection and holding capillaries (BioMedical Instruments, Germany).
10. Borosilicate glass capillaries with inner filament (GC100TF, Clark Capillaries, Harvard Apparatus Ltd., www.clark.mcmail.com).
11. Transfer, holding, and injection pipets (BioMedical Instruments).
12. Horizontal glass capillary puller (Sutter Instruments, model P97 or P87).
13. Inverted microscope (e.g., Zeiss Axiovert 10 or 200) and mechanic micromanipulators (Leica, cat. no. 11520137 and 11520138).
14. Control units for micromanipulation (Eppendorf CellTram Oil).
15. Microloader (Eppendorf, cat. no. 5242956.003).
16. Injector (Eppendorf FemtoJet) or 50-mL Falcon syringe connected to tubing.
17. Flexible silicon tubing.
18. Fine straight scissors and anatomical forceps.
19. Ketamine hydrochloride (e.g., Ketasol-100, 100 mg/mL, E. Gräub AG, Berne, Switzerland) and xylazine (e.g., Rompun 2% solution, Bayer). For working solution (kept at 4°C): 1 mL ketasol-100 and 0.8 mL rompun, made to 11 mL with double-distilled water or PBS. Inject intraperitoneally 0.1 mL/10 g body weight.
20. Wound clip (Aesculap BN507) with applicator (Aesculap BN731) or suture needles.
21. Stereomicroscope with cold light lamp (e.g., Zeiss, cat. no. KL 1500).
22. Humidified 5% CO_2 incubator at 37°C (e.g., Labotect).
23. Blastocyst cooling station (BioMedical Instruments).

2.4. Generation of Tissue-Specific Transgenic Lines

1. 4% paraformaldehyde (PFA) in 1X PBS. Adjust to pH 7.2. Make 5-mL aliquots and store at –20°C. Note that thawed vials should not be refrozen.
2. 20 mg/mL X-gal in dimethylformamide (Sigma, cat. no. D8654); store at –20°C in the dark.
3. 0.1 M potassium ferricyanide; keep at 4°C in the dark.
4. 0.1 M potassium ferrocyanide; keep at 4°C in the dark.

5. Detergent wash: 0.1 M sodium phosphate buffer (pH 7.4; it is important that it is prepared from 1 M stock made of 77.4 mL 1.0 M Na$_2$HPO$_4$ and 22.6 mL 1 M NaH$_2$PO$_4$), 2 mM MgCl$_2$, 0.01% desoxycholate, and 0.02% NP40. Detergent wash is prepared as a 2X solution and diluted 1:1 with distilled water (dH$_2$O) before use.

6. Staining solution: 1X detergent wash (*see* **step 5**), 3.33 mM potassium ferricyanide, 3.33 mM potassium ferrocyanide, 0.66 mg/mL X-gal, and 20 mM Tris-HCl at pH 7.0.

7. Plasmid Midi kit (Qiagen).

8. Gel extraction kit (QIAquick, Qiagen).

9. Microinjection buffer: 5–10 mM Tris-HCl at pH 7.4, 0.1 mM EDTA (commercially available from Specialty Media, MR-095-F; in Europe, Metachem Diagnostics, http://www.metachem.co.uk/SMedia.htm).

10. 0.22-µm filters (Millipore Millex-GV4, cat. no. SLGVL040S/SLGV004SL).

11. 50 mM NaOH.

12. 1 M Tris-HCl (pH 8.0).

13. Cre-primer: number 1: 5'-cctggaaaatgcttctgtccg-3', T_m = 59.8°C; number 2: 5'-cagggtgttataagcaatccc-3', T_m = 57.9°C.

14. DMSO.

15. Taq polymerase (e.g., Invitrogen) with corresponding 10X buffer.

16. PCR cycler (e.g., Biometra, PerkinElmer).

3. Methods

The methods described outline the construction of the conditional gene targeting vector (**Subheading 3.1.**), transfection of mouse ES cells (**Subheading 3.2.**), screening of targeted ES cell clones (**Subheading 3.3.**), generation of germline chimeras (**Subheading 3.4.**), and generation of tissue-specific knockout mice (**Subheading 3.5.**).

3.1. Construction of the Conditional Gene-Targeting Vector

A replacement-type construct is most commonly used. Such a targeting vector contains two homologous regions, located on either side of a mutation. Homologous recombination occurs by a double crossover event that replaces the target-gene sequences with the replacement-construct sequences. Although the basic technique for generating a targeted gene locus is also applied for conditional gene targeting, specific differences have to be taken into consideration because of the presence of loxP sites that flank the "vital region" of the gene locus or the possibility to eliminate the selection markers by Cre or Flp recombinases. The construction of the gene-targeting vector is described in **Subheadings 3.1.1.** and **3.1.2.** This includes the vector design for ES cell-based transgenesis (**Subheading 3.1.1.**) and test transformations to verify the functionality of loxP and frt sites (**Subheading 3.1.2.**).

3.1.1. Vector Design for ES Cell-Based Transgenesis

The targeting vector for the *Scnn1a* (α ENaC) gene locus contained 5'- and 3'-homologous sequences of the endogenous *Scnn1a* gene locus, a region containing an exon that encodes an essential region for gene function (exon 1, the vital region) flanked by loxP sites and a neomycin resistance marker (*neo*), followed by a third loxP site (*see* **Note 1**). In addition, it contained a negative selectable marker (herpex simplex virus thymidine kinase [HSV-*Tk*] gene) positioned outside the area of genomic homology and removed when homologous recombination occurred *(6)*.

1. In general, sequence information should be obtained on the genomic gene locus (National Institutes of Health or EMBL database) and from the isolated DNA source (e.g., phage, cosmid, BAC DNA) (*see* **Note 2**).
2. Define the boundaries of your targeting construct to flank the vital region by 5–8 kb of genomic sequence (*see* **Note 3**).
3. Design a convenient Southern blot strategy based on sequence or mapping information that allows recognition of homologous recombination from both 5' and 3' of the integration site using single-copy probes outside the targeting vector.
4. Choose loxP- and flp-containing vectors as well as the selection marker cassettes (e.g., neomycin resistance gene, HSV thymidine kinase gene) for cloning your target region (*see* **Note 4**).
5. DNA manipulations are performed by standard procedures.

3.1.2. Test Transformations

The recombination competence of Cre and Flp recombination target sites is tested by transformation into modified *E. coli* strains (294-Cre, 294-Flp *[12]*). Plasmids with recombinase recognition targets should be completely recombined, thus excising the DNA sequence between the two loxP and the two flp sites, respectively.

1. Transform plasmid into the 294-Cre or 294-Flp *E. coli* strains (*see* **Note 5**).
2. Grow overnight at 37°C.
3. Isolate plasmid DNA and assess recombination by restriction digestion.

3.2. Transfection of ES Cells

Mouse ES cells are electroporated with the targeting construct and subjected to double selection to enrich for recombinant ES cell clones. Approximately 1 in 10^4 transfected cells will stably integrate the DNA (the efficiency may vary depending on the ES cell line). Therefore, a dominant selectable marker is used to allow isolation of stable transfectants. Generally, this is based on positive selection of a drug resistance gene (e.g., *neomycin*) using drugs (e.g., G418), thus eliminating cells that have not stably incorporated the targeting vector.

However, the targeting construct will integrate rather randomly into the genome. Therefore, to further enrich for homologous recombinants, a negative selectable marker (e.g., HSV-*Tk*) is included in the construct, outside the region of homology to the target gene locus. In the presence of the HSV-*Tk* gene, cells become sensitive, for instance, to gancyclovir and die because sequences outside the homologous regions are lost during recombination but not in random integrants (*see* **Note 6**). The steps in this process thus involve the electroporation of mouse ES cells with the targeting construct (**Subheading 3.2.1.**), followed by selection and screening of targeted ES cell clones (**Subheading 3.2.2.**) and amplification and freezing procedures of single ES cell clones (**Subheading 3.2.3.**).

3.2.1. Electroporation of ES Cells

1. Start with an exponentially growing culture of early-passage ES cells. Most ES cells are maintained in coculture with irradiated embryonic mouse fibroblasts (feeder cells *[13]*) that should be thawed at least 1 d prior to thawing ES cells. Thaw an aliquot of ES cells into a T25 flask containing feeder cells. To obtain an exponentially growing culture of early-passage ES cells, pass cells at least twice (every 2–3 d, 1:3 to 1:4, with daily medium change). Change medium 3–4 h prior to electroporation. For trypsinization, rinse the cells twice with PBS (without Ca^{2+}/Mg^{2+}), add a small volume of trypsin (3 min at 37°C), rapidly add complete ES cell medium (DMEM, penicillin/streptomycin, β-mercaptoethanol, 15% heat-inactivated FCS serum) to inhibit trypsin action, and pipet the cells gently up and down to produce a single-cell suspension. Centrifuge (1200 rpm for 5 min) and resuspend cells in PBS (without Ca^{2+}/Mg^{2+}) at a density of 3×10^7 cells/mL.
2. For each electroporation, place 1.5×10^7 cells in 0.8 mL into a sterile cuvet 0.4 cm wide, add 20 µL linearized DNA (1 µg/µL), mix, and let the mixture incubate at room temperature for 5 min.
3. Mix again and apply a single 250-V pulse (1 ms, LV).
4. Allow cells to recover 5 min at room temperature, then transfer into 50-mL prewarmed ES cell medium and plate at a density of about 5×10^6 cells onto 10-cm dishes containing irradiated embryonic fibroblasts. One six-well dish is fed with nonselective medium to determine the percentage of stably transfected cells, and one 10-cm dish only contains G418 to calculate the enrichment in double-selected dishes. Generally, we perform two independent electroporations, which are then pooled.

3.2.2. Selection and Screening of Targeted Clones

Resistant colonies are screened by PCR or Southern blot analysis to detect ES cell colonies in which the DNA is recombined correctly.

1. Change medium 24 h following electroporation and start selection with G418 (400 µg/mL, single positive selection). Positive-negative selection is started 1 d

later using G418 and gancyclovir (2 μ*M*) to enrich for cells that contain a functional *neo^R* gene but have lost the *HSV-Tk* gene as a result of homologous recombination. Leave one separate 10-cm dish (only G418) to calculate the apparent enrichment.

2. Change medium every day (20 mL/dish) for 5 d consecutively, then stop the gancyclovir treatment and continue with the G418 treatment (15 mL/dish; positive selection).
3. Following culturing of transfected cells in selective medium, colonies become clearly visible after 8–10 d. Individual colonies can be isolated and expanded in 96-well plates to perform replica plating and screening.
4. The day before picking, prepare 96-well plates with irradiated embryonic fibroblasts.
5. Using an inverted microscope (placed under the hood), pick colonies with a finely drawn Pasteur pipet or glass capillary. It is important to pick single colonies of identical size. Generally, we pick during 2–4 d to obtain six 96-well plates per electroporation (*see* **Note 7**).

3.2.3. Amplification and Freezing Procedures

3.2.3.1. AMPLIFICATION

1. At 3–5 d following isolation of individual colonies, make three replica plates (three 96-well plates) and freeze two plates after 2–3 d of further culturing. Trypsinize the third plate and passage 1:3 into 96-well (flat-bottom) plates covered with gelatin. These plates will be used for DNA isolation (*see* **Note 8**).
2. To trypsinize 96-well plates, rinse twice with PBS (without Ca^{2+}/Mg^{2+} using an eight-channel aspirator/multichannel pipet), add 25 μL trypsin/EDTA solution, leave 3–4 min in the incubator, and resuspend the cells in 50–100 μL complete ES cell medium.

3.2.3.2. FREEZING PROCEDURES

1. For freezing of 96-well plates, use special V-shaped 96-well plates. Distribute 75 μL 2X freezing medium (70% FCS, 10% complete ES medium, 20% DMSO).
2. Resuspend trypsinized cells in 50 μL complete ES cell medium and transfer into individual wells while mixing using the multichannel pipet. Seal with parafilm, then cover with a paper towel and freeze at –80°C in suitable Styrofoam boxes.

3.3. Screening of Targeted ES Cells

The steps described in **Subheadings 3.3.1.–3.3.3.** outline the procedures for screening targeted ES cell clones by PCR and Southern blot analyses starting from 96-well tissue culture dishes *(14)*.

1. Highly confluent cells are rinsed twice with PBS. Add 50 μL lysis buffer and 0.5 mg/mL proteinase K per well. Incubate the plates in a humid chamber overnight at 56°C.

2. The next day, carefully add 100 μL NaCl/EtOH (150 μL 5 M NaCl and 10 mL cold ethanol) per well using a multichannel pipet. Leave at room temperature for 30 min to allow precipitation of nucleic acids.
3. Spin down 5 min at 850g and remove NaCl/EtOH by inverting the dish carefully.
4. Wash three times with 150 μL 70% EtOH. After the final wash, invert the dish carefully and dry for 10–15 min. Resuspend in 30 μL TE for Southern blot analysis or in 50 μL TE for PCR analysis. Incubate the plates for several hours at 37°C (e.g., in an incubator). They can then be used for further analyses or stored at –20°C.

3.3.1. Screening by PCR

All PCR screening must be performed under stringent conditions to avoid contamination and false positives. This is done under a hood or on a clean separate bench using filter tips and separate pipets. In general, correct recombination is identified using one primer in a marker gene (e.g., neo^R cassette) or near the loxP site and the second corresponding primer outside the targeting vector.

1. Make the PCR in 2.5 μL 10X PCR buffer (Mg^{2+} free), 2.5 μL 25 mM MgCl$_2$, 4 μL dNTP (2.5 mM each), 0.25 μL of each primer (100 ng/μL), 0.25 μL TaKaRa La Taq polymerase (5 U/μL), 14.25 μL dH$_2$O, and 1 μL DNA.
2. Amplification is done using the following conditions (*see* **Note 9**): 3 min at 95°C; 30 s at 95°C, 30 s at 64°C, 1 min at 68°C 20 times (decline 0.5°C each cycle); 30 s at 95°C, 30 s at 54°C, 1 min/1000 bp at 68°C) for 30 cycles; and 5 min at 72°C; store at 4°C.
3. Load with a multichannel pipet onto a 0.8% agarose gel (TBE) to visualize amplified PCR products.

3.3.2. Screening by Southern Blot Analysis

1. Following extraction, genomic DNA is directly digested in the well. The plates are subjected to DNA digestion ("mini" Southern blot analysis) by adding 10 μL digestion mix (4 μL [10 U] restriction enzyme, 4 μL 10X reaction buffer, 0.5 μL 100 mM spermidine, 1.5 μL ddH$_2$O). Mix the contents with the tip of the multichannel pipet and incubate the dish in a humidified incubator at 37°C overnight (final volume 40 μL; *see* **Note 10**).
2. Take an aliquot (2 μL digestion reaction) and check complete digestion on a TBE test gel containing E+Br. For Southern blot analysis, take the resting 38 μL digested DNA solution add 8 μL TAE electrophoresis loading buffer and run an 0.8% TAE gel without EtBr.
3. Treat gel for 10 min in 0.25 N HCl, 30 min in 0.4 M NaOH, and 20 min in neutralizing solution.
4. Wet a nylon membrane in double-distilled water and 2X SSC and transfer overnight using paper towels and 20X SSC buffer via capillary forces ("classical" Southern blot transfer).

5. The next day, wash membrane briefly in 2X SSC, let air-dry, and crosslink (autocrosslink, 1200 J).
6. Rehydrate the membrane in 25 mM phosphate buffer/1 mM EDTA. Prepare ^{32}P-labeled probes using standard procedures.
7. Following pre- and hybridization of the membranes overnight at 65°C, wash filter and expose to autoradiography or phosphoimager.

3.3.3. Amplification of Targeted ES Cell Clones

Following identification of recombined clones, go back to the frozen 96-well plate, select 4–10 positive clones, amplify, and freeze aliquots. At this step, targeted clones should be confirmed by detailed DNA analysis and be tested for mycoplasma and chromosome instability *(15)*.

3.3.3.1. KARYOTYPE ANALYSIS

Cells can be arrested in metaphase by the addition of colcemid (colchicin) to the culture medium. For best results, use an exponentially dividing population of ES cells.

1. Take a confluent six-well dish and split 1:3 onto feeder-free gelatinized six-well dishes.
2. The next day, replace medium with 3 mL medium containing 0.06 μg/mL colchicin (demecolcin). Incubate 3–4 h at 37°C.
3. Collect cells by standard procedures (*see also* **Subheading 3.2.1.**), transfer to a 10-mL centrifuge tube, and pellet the cells.
4. Aspirate supernatant but leave about 0.3 mL. Add 10 vol of 0.56% KCl solution and resuspend cells carefully. Incubate 5–10 min to allow the cells to swell.
5. Pellet the cells by gentle centrifugation (5 min at 50g). Remove supernatant and mix.
6. Using a Pasteur pipet, add dropwise 5 mL ice-cold fixative (3:1 vol of absolute methanol to glacial acetic acid, freshly prepared) while vortexing or shaking the tube to ensure that the cells do not form large clumps. Incubate for 10 min at room temperature.
7. Pellet cells gently (50 rpm, 5 min) and resuspend in 2 mL fixative.
8. Release single drops of cell suspension onto clean glass microscope slides (cleaned with 70% ethanol) from a height of about 40 cm, blow across the surface, and let air-dry. Stain with Giemsa solution (about 10 min), rinse with water, let air-dry, put under cover slip, and count chromosomes.

3.4. Generation of Germline Chimeras

Once identified and confirmed using Southern blot strategy, targeted ES cell clones are injected into host blastocysts to give rise to chimeras that might transmit the modified ES cell genome to offspring. For ES cell lines derived from 129 strains, C57BL/6 host blastocysts have proven to be suitable because the coat color of chimeras is easily distinguishable (as being agouti on a nonagouti background) and because ES cells have an optimal developmental advantage, which allows the generation of highly chimeric animals. The handling and general conditions of both embryo donor and recipient mice are of major importance for the successful outcome of microinjection experiments. For further details on optimal husbandry, refer to **ref. *13***.

Handling and culture of embryos, the microinjection process, and the final transfer of the embryos into the uterus/oviduct of pseudopregnant females require experience and optimized culture conditions. The microinjection process itself requires expensive and specialized equipment. Therefore, these experiments should not be undertaken without careful consideration of required sources, tools, and training of the technical staff. Many universities and research institutes have core facilities that perform these tasks take contract work from outside groups.

Described in **Subheadings 3.4.1.** and **3.4.2.** are the steps that are normally utilized to generate germline chimeras from preparation to injection of ES cells into blastocyst, followed by subsequent transfer into foster mice.

3.4.1. Preparation of ES Cells for Injection

It is important to reduce the time during which ES cells are in culture between selection and injection into blastocysts.

1. For blastocyst injection, thaw a vial of ES cells in three six-well dishes covered with irradiated fibroblasts. At 1 d prior to injection, select a 60–70% confluent dish, which is then trypsinized and split 1:2 to 1:3.
2. Immediately before injection, cells are again trypsinized and resuspended in about 1 mL complete ES cell culture medium. Put on ice.

3.4.2. Injection of ES Cells Into Blastocysts and Transfer of Blastocysts Into Foster Mothers

The efficiency of the blastocyst injection is largely influenced by the number of the blastocysts available, the quality of the used ES cell line and the manual skills of the operator. The embryos are easiest handled (collection, sorting, transfer) by careful mouth pipetting using finely drawn Pasteur pipets or transfer pipets from commercial suppliers. Blastocyst injection is still one of the most commonly used methods for generating chimeras. If high chimerism has been achieved in the mice, then the likelihood is great that gene-modified ES cells will contribute to the germline, essentially via the male germ cells. By breeding such chimeras to wild-type mice, it is thus possible to establish mouse lines carrying the targeted allele.

For microinjection, an inverted microscope with phase contrast optics, micromanipulators, and low (×5 or ×10) and high magnification objectives (×20 or ×32) is needed. If oocyte injection will also be performed, then Nomarski optics are preferable. Two micromanipulators are needed for the fine movement of the holding and injection capillaries. The holding and injection capillaries are connected by tubing to control units. We use the following conditions for blastocyst injection:

1. 8- to 14-wk-old C57BL/6N females are spontaneously mated with C57BL/6N males, and plugged females are separated in the morning (d 0.5 = d 3 before injection). The mice are maintained on a 14/10 h light/dark cycle.
2. The next day, 8- to 14-wk-old females of an outbred strain (NMRI) are spontaneously mated with vasectomized NMRI males to obtain pseudopregnant females (plug positive the next morning, d 2 before injection). Those males can be ordered from commercial suppliers (e.g., from Janvier, France).
3. At d 3.5, plugged C57BL/6N females are dissected, and blastocysts are flushed using M2 or blastocyst injection medium (*see* **step 4**). They are incubated (for 1–2 h) in ES cell medium (DMEM without β-mercaptoethanol, 5% CO_2, 37°C) until injection.
4. 10–15 blastocysts are transferred to the injection chamber (plastic or glass frame mounted with paraffin on a glass slide). A drop of blastocyst injection medium (ES cell medium containing 7.5% FCS and 10 mM HEPES) (*see* **Note 11**) covered with mineral or silicon oil (*see* **Subheading 2.3.**, **steps 5 and 6**) is put onto the blastocyst cooling station (set at around 12°C). Blastocyst injection is performed as described *(13)* using 5–15 ES cells in dependence of the ES cell line.
5. To ensure a litter size of four or more pups, around 15 blastocysts are transferred into the uterus (both sides) of a pseudopregnant female (*see* **Note 12**).

3.5. Generation of Conditional Knockout Mice

Mice expressing Cre recombinase are required to remove a floxed gene in a tissue-specific manner *(3,16)*. Such mice are essentially generated using transgenic constructs. However, it is equally feasible to introduce a Cre recombinase gene into a gene locus of interest ("knock-in") to obtain Cre recombinase expression that closely mimicks regulation in time and space of the endogenous gene *(17)*. A further refinement (not discussed here) is the use of an inducible Cre recombinase with activity that is dependent on, for example, tamoxifen *(16,18)*. The following steps include the design of the construct (**Subheading 3.5.1.**), the generation of transgenic mice by pronuclear injection (**Subheading 3.5.2.**), the testing of Cre recombinase activity in the transgenic strains (**Subheading 3.5.3.**), and the breeding to generate conditional knockout mice (**Subheading 3.5.4.**).

3.5.1. Design of Constructs

The Cre coding sequence is obtained from pBS185 *(19)* or pMC-*Cre* *(20)*, either with or without the polyA sequences. Because the polyA sequences in these plasmids have no introns, it is advisable to add an SV40 splice and polyadenylation site. To target Cre expression, an established promoter, already proven in transgenic experiments, is preferable. It is feasible to test the functionality of the Cre-expressing transgene construct in cell culture.

The transgene construct is cotransfected with a reporter plasmid allowing, for example, Cre recombinase-mediated activation of green fluorescent protein expression *(17)*. This allows assessment of the integrity of the construct before generating transgenic mice. Finally, the promoter is cloned in front of the *Cre*-polyA, keeping unique restriction enzyme sites on both ends of the insert. Using homologous recombination in ES cells, a *Cre*-polyA cassette can be inserted into a given locus (knock-in). It is highly recommended to remove the neomycin cassette (use frt sites surrounding the neo^R cassette) in mice because the presence of the Pgk promoter or the *neomycin* gene might interfere with expression from the endogenous promoter once the mice have been obtained. However, a *Cre* knock-in approach can only be used if the deletion of the gene is harmless and without phenotype in heterozygous condition. Note that this is often not known *a priori*.

As an alternative to standard transgenic constructs, the Cre recombinase gene might be placed under control of a gene promoter in the context of a BAC using homologous recombination in bacteria *(21,22)*. Compared to conventional minigene constructs, BAC transgenes should be more reliable in expression pattern and level, mainly because of their size of 150–200 kb, thus encompassing all regulatory elements, even those potentially located in introns. BACs can be best injected when uncut and supercoiled. However, BAC vectors often contain loxP sites, which have to be removed by homologous recombination. Alternatively, the insert can be separated from the vector and recovered following pulse field electrophoresis.

3.5.2. Generation of Transgenic Mice

The generation of transgenic mice follows standard procedures *(13)*. We use fertilized oocytes from B6D2F1 male and female matings. Next described are the microinjection of DNA into the pronucleus (**Subheading 3.5.2.1.**), the identification of transgenic mice (**Subheading 3.5.2.2.**), and the testing of Cre recombinase by breeding to reporter mice (**Subheading 3.5.2.3.**).

3.5.2.1. Microinjection of DNA Into the Pronucleus

The insert is retrieved from vector sequences, purified, and diluted into injection buffer.

1. Purify recombinant plasmid using, for example, a Qiagen kit (Qiagen, Plasmid Midi kit).
2. Digest 10–20 µg of plasmid with restriction enzyme to remove vector sequences. Run minigel to check complete digestion (*see* **Note 13**).
3. Separate insert from vector on agarose gel. Cut out the band (minimize ultraviolet light exposure, 420-nm wavelength) and transfer gel slice into Eppendorf tube.
4. Purify insert by QIAquick Gel Extraction kit (*see* **Note 14**).
5. Resuspend recovered DNA in filtered microinjection buffer (*see* **Note 15**).
6. Determine DNA concentration of sample and run a minigel with molecular weight markers of known concentrations. Check DNA for purity, correct size, and absence of smearing or significant amount of uncut plasmid vector sequence. Adjust final DNA concentration of sample to 1–5 ng/µL by comparison to previously injected constructs. Filter DNA through 0.22-µm filters and store two or three (20- to 50-µL) aliquots in dust-free Eppendorf tubes at –20°C (aliquots can be reused several times).
7. B6D2F1 females (3–4 wk old or older than 8 wk) are superovulated using 5 IU pregnant mare's serum gonadotropin (2–4 PM) and 5 IU human chorionic gonadotropin (10–11 AM 2 d later) and mated in the late afternoon with B6D2F1 males (one female per male). Check for plugs the following morning. Light period is 6 AM to 8 PM.
8. Before injection, DNA is centrifuged for 10 min to pellet remaining dust particles. A small volume of DNA solution is introduced into the injection pipet using a microloader. Absence of air bubbles in the tip is controlled under the stereomicroscope, and the pipet is then connected to microinjection setup.
9. Fertilized oocytes are recovered at 1–2 PM in M2 medium and injected. They are transferred the day of injection (as one-cell embryos) into pseudopregnant females (outbred strain NMRI > 30 g). Operated females are housed two mice per cage.
10. Offspring is genotyped at an age of 3 wk for the presence of Cre recombinase using either Cre-specific primers or transgene-specific primers located in Cre recombinase (antisense primer) and promoter (specific primer; *see also* **Subheading 2.4.**, **item 12**).

3.5.2.2. IDENTIFICATION OF TRANSGENIC MICE

1. Take tail biopsies (2 mm) and transfer to Eppendorf tubes. Clean the scissors or scalpel between every mouse. Tail biopsies can be stored at –80°C until further analysis.
2. Add 600 µL 50 m*M* NaOH and heat to 95°C for 10 min. To avoid opening, take a needle and perforate the lid.
3. Vortex and add 50 µL 1 *M* Tris-HCl (pH 8.0).
4. Centrifuge 10 min in desk centrifuge (maximal speed) and transfer the supernatant to a new tube.
5. Make the PCR in the following reaction: 2.5 µL 10X PCR buffer, 2.5 µL DMSO, 1.75 µL dNTP (25 m*M*), 0.5 µL of each primer (0.1 m*M*), 0.25 µL of Taq

polymerase (5 U/μL), 16 μL ddH$_2$O, 1 μL DNA. The PCR is done under following conditions: 5 min at 94°C, 30 cycles of 30 s at 94°C, 1 min at 57°C, 1 min at 72°C, followed by 10 min at 72°C.

3.5.2.3. Testing Cre Recombinase by Breeding to Reporter Mice

Transgenic founder mice are first bred to C57BL/6 mice (or other strains) to propagate the line (*see* **Note 16**) and second to reporter mouse strains to monitor Cre recombinase activity in vivo (*see* **Note 17**). In contrast to standard transgenic experiments, Cre gene expression or protein expression is normally not analyzed and instead is replaced by direct evaluation of Cre activity. The most used reporter strain is Rosa26R, for which ubiquitous expression of a lacZ gene is blocked by a STOP cassette flanked by loxP sites *(23)*. Cre activity leads to removal of the STOP, and lacZ expression is effective. In dependence of the promoter, Cre activity is monitored in embryos (*see* lacZ protocol below) or on cryosections of adult tissue (*see* **Note 18**). Further details on the protocol can be found in **refs.** *13* and *24*:

1. Dissect E13.5 embryos from the pregnant female.
2. Remove all traces of blood in the Petri dish with PBS.
3. Fix embryos in PFA (4%) at 4°C (on ice) for about 30–60 min (*see* **Note 19**).
4. Wash embryos twice briefly in PBS and then in 1X detergent wash twice for 45 min at room temperature.
5. Prepare 5 mL fresh staining solution (sufficient for 10 embryos) and stain embryos at 37°C in the dark.
6. Stop staining (after 4–24 h) when signal is strong or when background staining appears.
7. Wash embryos briefly in PBS and store in PFA at 4°C.
8. Photograph embryos and process to histology, if required.

3.5.3. Generation of Conditional Knockout Mouse

Cre transgenic mice are bred to mice carrying the floxed gene locus of interest (*see* **Note 20**). Most often, two rounds of breedings are sufficient to obtain homozygous floxed mice carrying the Cre transgene. Recombination efficiency is best measured by isolating a pure population of cells followed by Southern blot analysis (or quantitative PCR) (*see* **Note 21**). Failure to obtain transgenic mice on a homozygous "floxed" background is mostly because of recombination and a deleterious effect of the gene during embryonic development. Rarely, it can also happen that the transgene is inserted very closely to the floxed gene locus, thus rendering crossovers very unlikely. In that case, either try a large number of litters or use another independent Cre transgenic line.

4. Notes

1. An alternative approach is based on Flp-mediated recombination *(3,25,26)*. The selection marker *neo* is flanked by two frt sites ("flirted") and can be deleted using

Flp recombinase. This step is performed in vivo using transgenic mice expressing Flp recombinase to obtain offspring without the *neo* gene or in vitro in targeted ES cells by transient expression of eukaryotic FLPe expression vector *(4,5)*.

2. Genomic fragments of the mouse *Scnn1a* gene locus were isolated from a 129/Sv mouse genomic library (Stratagene). As an alternative to the BAC clones, genomic DNA can be ordered (e.g., from Taconic) and isolated using long-range, high-fidelity PCR. It is recommended to sequence all exon sequences.

3. For the *Scnn1a* gene-targeting vector, fragments were subcloned into the *Pme*I, *Asc*I, and *Sal*I sites of a modified loxP-targeting vector containing three loxP sites *(27)*. For 5'-homology, a 4.3-kb fragment was placed upstream of the 5' loxP site. For the vital region, a 1.5-kb fragment containing coding sequences (exon 1) and upstream sequences (position –912 relative to the ATG) and downstream sequences (partial first intron, position +609) was amplified by PCR, sequenced, and subcloned into the targeting vector using primers containing *Asc*I sites. For the 3' homology, a 1.9-kb fragment was introduced downstream of the 3' loxP site.

4. We used a neo/TK vector containing three loxP sites, kindly provided by F. Radtke (Ludwig Institute, Epalinges) *(27)*. For another conditional knockout project *(26)*, this vector was combined with a vector containing a flirted neo cassette (K13pATRTneolox antisense, kindly provided by A. Trumpp, ISREC, Epalinges; *28*) (*see also* **Subheading 2.1., step 17**).

5. Because of problems of DNA instability, we have used a modified protocol for competent 294-Cre *E. coli* and transformation. Bacteria are grown to an OD_{600} of 0.3–0.6, pelleted (1000*g*, 10 min), and resuspended in transformation storage buffer (LB, 10% p/v PEG6000, 50 m*M* $MgCl_2$ at pH 6.5; supplement with 5% DMSO). Plasmid DNA is dissolved in 100 µL 0.5 *M* KCl, 0.15 *M* $CaCl_2$, and 0.25 *M* $MgCl_2$ and mixed with 100 µL competent bacteria (20 min on ice, 10 min room temperature). 500 µL LB is added, incubated at 37°C on a shaker (30–60 min), and plated.

6. A considerable percentage of ES cell clones will carry an inactive HSV-*Tk* gene even when no homologous recombination has occurred. Thus, gancyclovir-resistant clones have to be screened for recombination.

7. We normally use small capillaries to mechanically separate the colonies in small fragments and aspirate and transfer those into individual 96-well vials. Other methods use picking of single colonies followed by trypsinization *(13)*.

8. Pretreatment with gelatin: Use flat-bottom 96-well plates. Add 40–50 µL 0.1% gelatin solution for 5 min (*see also* **Subheading 2.2., item 13**). Remove gelatin solution and replace by complete ES cell culture medium. It is not necessary to dry the plates before adding the medium.

9. Generally, we start the annealing temperature about 10°C higher than the calculated temperature. The temperature is decreased by 0.5°C during each of the first 20 cycles.

10. As an alternative, we use a simple incubator and put the dish in a closed box (e.g., Tupperware) with humidified paper towels.

11. As an alternative to a cooling chamber, blastocysts are shortly incubated in cold

(4–10°C) injection medium before transfer to the injection chamber.

12. As an alternative, blastocysts can be transferred into the oviduct of d 0.5 pseudo-pregnant females. Correspondingly, females will give birth only after 19–20 d.

13. Because it is often difficult to distinguish linearized plasmid from insert in large constructs, this should be checked using appropriate restriction enzyme digestions.

14. The quality of DNA for microinjection is essential to the success of transgenic experiments. DNA that is not purified properly will make the injections difficult or reduce survival. Any traces of phenol, ethanol, salts, or enzymes may be toxic for embryos. It is also important to remove all particles that could clog the injection needles. It is recommended to filter all solutions.

15. Use powder-free gloves (or wash gloves with water) to avoid potential clogging of injection needles.

16. Cre overexpression can lead to recombination *(29)*, probably because of the existence of endogenous loxP-like sequences in the genome *(30)*. It is thus highly recommended to test Cre transgenic mice for potential unwanted side effects.

17. It should be noted that some tissue-specific Cre lines will result in germline expression of the Cre recombinase. For example, breeding of a Dct::Cre-transgenic male that equally carried the Rosa26 reporter transgene resulted in completely "blue" offspring *(17)*.

18. Most reporter strains (Rosa26R, Z/EG, Z/AP) do function well during embryonic development *(23,31,32)*. However, not all tissues will express from the Rosa26 promoter or the chicken β-actin promoter during adulthood. Thus, different reporter strains have to be tested, and the reporter strain might be preselected using an ubiquitous expression (e.g., following breeding to a germline deleter strain).

19. Many protocols exist, and some use lower concentrations of PFA or use glutaraldehyde instead. In any case, fixation times might vary and should be tested because they will depend on the embryo size, the site of expression, and the strength of the promoter. Staining in external sites (skin, ear, eye) will certainly require less fixation then internal organs (lung, kidney). It might also be envisaged to cut embryos in half before starting the procedure *(13)*.

20. Cre might exert a certain "toxicity" (*see also* **Note 16**). Thus, a very first control group of mice should contain Cre transgenic animals, which are wild-type or heterozygous mutant.

21. Recombination efficiencies can vary between different target genes *(33)*. When using a Cre transgenic line, a certain locus might lead to 50% efficiency, another to all cells recombining in both alleles. To increase efficiency, it is feasible to use a knockout allele (e.g., by breeding a Cre transgenic lox/Δ male with lox/lox females) if there is no heterozygous effect in the tissue of interest.

Acknowledgments

Thanks are due to Andrea Schmidt, Anita Rossier, Christelle Richard, and Barbara Canepa for contributing to blastocyst and pronuclear injections. Work in the laboratory of F. B. is supported by grants from the Swiss Cancer League, the Swiss National Science Foundation, and the National Center of Competence in Research Molecular Oncology (SNF); work in the laboratory of E. H. is supported by the Swiss National Science Foundation. The Transgenic Animal Facility (TAF) is supported by the Faculty of Biology and Medicine and University Hospitals (CHUV) of Lausanne.

References

1. Sternberg, N. and Hamilton, D. (1981) Bacteriophage P1 site-specific recombination. I. Recombination between loxP sites. *J. Mol. Biol.* **150,** 467–468.
2. Sternberg, N., Hamilton, D., and Hoess, R. (1981) Bacteriophage P1 site-specific recombination. II. Recombination between loxP and the bacterial chromosome. *J. Mol. Biol.* **150,** 478–507.
3. Nagy, A. (2000) Cre recombinase: the universal reagent for genome tailoring. *Genesis* **26,** 99–109.
4. Dymecki, S. M. (1996) Flp recombinase promotes site-specific DNA recombination in embryonic stem cells and transgenic mice. *Proc. Natl. Acad. Sci. USA* **93,** 6191–6196.
5. Rodriguez, C. I., Buchholz, F., Galoway, J., et al. (2000) High-efficiency deleter mice show that FLPe is an alternative to Cre-loxP. *Nat. Genet.* **25,** 139–140.
6. Hummler, E., Mérillat, A.-M., Rubera, I., Rossier, B., and Beermann, F. (2002) Conditional gene targeting of the *Scnn1a* (alpha ENaC) gene locus. *Genesis* **32,** 169–172.
7. Rossier, B. C., Pradervand, S., Schild, L., and Hummler, E. (2002) Epithelial sodium channel and the control of sodium balance: interaction between genetic and environmental factors. *Annu. Rev. Physiol.* **64,** 877–897.
8. Barker, P. M., Ngugen, M. S., Gatzy, J. T., et al. (1998) Role of γENaC subunit in lung liquid clearance and electrolyte balance in newborn mice: insights into perinatal adaptation and pseudohypoaldosteronism. *J. Clin. Invest.* **102,** 1634–1640.
9. Hummler, E., Barker, P., Gatzy, J., et al. (1996) Early death due to defective neonatal lung liquid clearance in αENaC-deficient mice. *Nat. Genet.* **12,** 325–328.
10. McDonald, F. J., Yang, B., Hrstka, R. F., et al. (1999) Disruption of the β subunit of the epithelial Na⁺ channel in mice: hyperkalemia and neonatal death associated with a pseudohypoaldosteronism phenotype. *Proc. Natl. Acad. Sci. USA* **96,** 1727–1731.
11. Rubera, I., Loffing, J., Palmer, L. G., et al. (2003) Collecting duct specific gene activation of alphaENaC in the mouse kidney does not impair sodium and potassium balance. *J. Clin. Invest.* **112,** 554–565.
12. Buchholz, F., Angrand, P.-O., and Stewart, A. F. (1996) A simple assay to determine the functionality of Cre or FLP recombination targets in genomic manipulation constructs. *Nucleic Acids Res.* **24,** 3118–3119.
13. Nagy, A., Gertsenstein, M., Vintersten, K., and Behringer, R. (2003) *Manipulating the Mouse Embryo-A Laboratory Manual*, Cold Spring Harbor Laboratory Press, Cold Spring Harbor, NY.

14. Ramirez-Solis, R., Rivera-Perez, J., Wallace, J. D., Wims, M., Zheng, H., and Bradley, A. (1992) Genomic DNA microextraction: a method to screen numerous samples. *Anal. Biochem.* **2,** 331–335.
15. Robertson, E. J. (1987) *Teratocarcinomas and Embryonic Stem Cells-A Practical Approach,* IRL Press, Oxford, WA.
16. Kwan, K. (2002) Conditional alleles in mice: practical considerations for tissue-specific knock-outs. *Genesis* **32,** 49–62.
17. Guyonneau, L., Rossier, A., Richard, C., Hummler, E., and Beermann, F. (2002) Expression of Cre recombinase in pigment cells. *Pigment Cell Res.* **15,** 305–309.
18. Feil, R., Wagner, J., Metzger, D., and Chambon, P. (1997) Regulation of Cre recombinase activity by mutated estrogen receptor ligand-binding domains. *Biochem. Biophys. Res. Commun.* **237,** 752–757.
19. Sauer, B. (1993) Manipulation of transgenes by site-specific recombination: use of Cre recombinase. *Methods Enzymol.* **225,** 890–900.
20. Gu, H., Zou, Y., and Rajewsky, K. (1993) Independent control of immunoglobulin switch recombination at individual switch regions evidenced through Cre-loxP mediated gene targeting. *Cell* **73,** 1155–1164.
21. Lee, E. C., Yu, D., Martinez de Velasco, J., et al. (2001) A highly efficient *Escherichia coli*-based chromosome engineering system adapted for recombinogenic targeting and subcloning of BAC DNA. *Genomics* **73,** 56–65.
22. Sparwasser, T., Gong, S., Li, J., and Eberl, G. (2003) General method for the modification of different BAC types and the rapid generation of BAC transgenic mice. *Genesis* **38,** 39–50.
23. Soriano, P. (1999) Generalized lacZ expression with the ROSA26 Cre reporter strain. *Nat. Genet.* **21,** 70–71.
24. Schmidt, A., Tief, K., Foletti, A., et al. (1998) LacZ transgenic mice to monitor gene expression in embryo and adult. *Brain Res. Prot.* **3,** 54–60.
25. Meyers, E. N., Lewandoski, M., and Martin, G. R. (1998) An Fgf8 mutant allelic series generated by Cre- and Flp-mediated recombination. *Nat. Genet.* **18,** 136–141.
26. Rubera, I., Meier, E., Vuagniaux, G., et al. (2002) A conditional allele at the mouse channel activating protease 1 (Prss8) gene locus. *Genesis* **32,** 173–176.
27. Radtke, F., Wilson, A., Stark, B., et al. (1999) Deficient T cell fate specification in mice with an induced inactivation of Notch1. *Immunity* **10,** 547–558.
28. Trumpp, A., Refaeli, Y., Oskarsson, T., et al. (2001) c-myc regulates mammalian body size by controlling cell number but not cell size. *Nature* **414,** 768–773.
29. Schmidt, E., Taylor, D., Prigge, J., Barnett, S., and Capecchi, M. (2000) Illegitimate Cre-dependent chromosome rearrangements in transgenic mouse spermatids. *Proc. Natl. Acad. Sci. USA* **97,** 13,702–13,707.
30. Thyagarajan, B., Guimaraes, M., Groth, A., and Calos, M. (2000) Mammalian genomes contain active recombinase recognition sites. *Gene* **244,** 47–54.
31. Novak, A., Guo, C., Yang, W., Nagy, A., and Lobe, C. G. (2000) Z/EG, a double reporter mouse line that expresses enhanced green fluorescent protein upon Cre-mediated excision. *Genesis* **28,** 147–155.

32. Lobe, C. G., Koop, K. E., Kreppner, W., Lomeli, H., Gertsenstein, M., and Nagy, A. (1999) Z/AP, a double reporter for Cre-mediated recombination. *Dev. Biol.* **208,** 281–292.
33. Vooijs, M., Jonkers, J., and Berns, A. (2001) A highly efficient ligand-regulated Cre recombinase mouse line shows that loxP recombination is position dependent. *EMBO Rep.* **2,** 292–297.

Index